Stingless Bees – An Unexplored Pollinator in India

Dr Avinash Chauhan
Scientist
Department of Entomology
SASRD, Nagaland University
Medziphema, Nagaland 797 106

and

Dr Hemant Kumar Singh
Professor and Head
Department of Entomology
SASRD, Nagaland University
Medziphema, Nagaland 797 106

भाकृअनुप
ICAR

Published by

Directorate of Knowledge Management in Agriculture
Indian Council of Agricultural Research
Krishi Anusandhan Bhavan I, Pusa, New Delhi 110 012

PRINTED : March 2022

Project Director (DKMA) : Dr S.K. Malhotra

Incharge (English Editorial Unit) : Ravindra Verma
Associate : Vivek Chauhan

Incharge (Production Unit) : Punit Bhasin

ISBN : 978-81-7164-246-5

Price : ₹ 500/-

Published by Dr S.K. Malhotra, Project Director, Directorate of Knowledge Management in Agriculture, Indian Council of Agricultural Research, Pusa, New Delhi 110012; Lasertypeset at M/s Print-O-World, 2568, Mandir Lane, Shadipur, New Delhi 110 008, and printed at M/s Chandu Press, 469, Patparganj Industrial Estate, Delhi 110 092.

Preface

With the evolution, the nature started developing different creatures. Both plant and animal kingdom developed vastly into different species. Among all animal species, insects are more primitive and dominant animals on this earth. Insects are considered harmful to man usually but hardly 1% of insect species fall in the pest category. Benefits of insects in maintaining economy outweigh the injury inflicted. Some of the beneficial insects are silkworms, bees, lac insects, parasites, predators and other biological control agents. Honeybees are one of the few insects directly beneficial to man. In the animal kingdom they belong to: Phylum-Arthropoda, Class-Insecta, Order-Hymenoptera, Superfamily-Apoidea and Family-Apidae. Honey is highly valued food produced by honeybees and it is also used as medicine. In addition to honey, other products like bees wax, pollen, royal jelly and bee venom are also produced from bee colonies. However, more than the value of these hive products; bees perform an important task of nature by assisting in pollination of plants when collecting its food which consists of nectar and pollen. The science of beekeeping is known as apiculture.

We can learn a lot from these little wonderful creatures. Honeybees are admired for their industriousness, unity, self sacrifice, tolerance and division of labour. Even the most feared bee stings help in healing muscular pains, rheumatism, arthritis and reduction in cholesterol level. Bee keeping can be practised as an ideal hobby or taken up as a part-time or full-time business.

Primitive man used to rob bee colonies found in the cavities of hollow trees or on rocks and trees and this is still being followed by some tribes. Development of modern bee keeping has its origin between 1500 and 1851 when many attempts were made to domesticate bees in different types of hives but were not successful because bees attached their combs together as well as to the walls of hive and combs required had to be cut for honey. Discovery of the principle of bee space in 1851 by L L Langstroth in USA resulted in first truly movable frame hive. This bee space was 9.5 mm for *Apis mellifera*. This discovery was followed by subsequent innovations like comb foundation mill, honey extractor, smoker, etc., which helped in the development of modern bee keeping we see today.

With the increasing industrialization and modernization in agricultural practices, we realized less presence of bees. Different researchers come up with the inferences that bees are one of the most important factors for human sustenance on the planet. Famous scientist, Einstein once said, If there were no bees, humans will also get eliminated in few years. With time agriculturists also started realizing that without bees we cannot increase the crop production and productivity. The concept of pollination came into being, which can be described as an essential component of the ecosystem service that results in the increase in food security and improvement of livelihoods. Although maximum insects are pollinators but among them, honeybees, stingless bees, bumble bees, solitary bees and dipterans are few important pollinators known for their pollination efficiency. Honeybees are well explored from last 500 years but stingless bees are still the unexplored ones. This book has tried to explain the importance of these neglected stingless bee pollinators. The book is divided into two parts. Part one consist of 11 chapters describing the stingless bees, their diversity, nest architecture, floral resources, hive products and scientific management while the part two consists of 2 chapters dedicated to the development of stingless bee for pollination, their role in agriculture and future prospects.

We are indebted to the Vice Chancellor, Nagaland University, Prof. Pardeshi Lal, Pro-Vice Chancellor, School of Agricultural Sciences and Rural Development, Medziphema, Prof. Aleminla Ao, Dean School of Agricultural Sciences and Rural Development, Prof. Akali Sema, Project Coordinator, All India Co-ordinated Research Project on Honeybees & Pollinators, New Delhi, Dr Balraj Singh, Member (Plant Science), Agricultural Scientist Recruitment Board, New Delhi, Dr P K Chakrabarty, Prof & Head, Division of Entomology, IARI, New Delhi, Dr Debjani Dey, Scientist AICRP (HB&P) Dr Kumaranag KM, Ngukho, professional, AICRP (HB&P), Medziphema for their everwilling help and constant encouragement throughout the preparation of the book.

Avinash Chauhan
H K Singh

March 2021

Contents

CHAPTER 1

Bees and their Importance

Earth originated about 4.5 billion years ago and with time, life on earth started emerging. The plants and animals both originated and evolved at different levels and today what we see all around is the masterpiece from evolutionary work of billions of years. Among all animals, insect's evolution took place so diversely that half of animal biomass today is of insects. They have dominated the earth and excel all other organisms in diversity and magnitude. They developed the ability to live in all kinds of weather and climatic conditions. They are ubiquitous of all animals and found wherever life can exist from freezing temperatures of Poles to burning mercury of desserts. They vary in size from minute beetles to macroscopic moths. The insects learnt various ways to survival and evolution projected them with different appearances as beetles (hard body covering), moths and butterflies (camouflage and imitation), wasps and bees (with sting apparatus), flies (quick flying) and other group of insects with different protection tactics. Bees, ants and termites came to live in social groups with the advancement of characters like alternation of generation, division of labour and hormonal control of social group called colonies.

The bees are highly developed social insects which signify the presence of a large clan of individuals with certain characters, castes and part that cohabit in a common nest to complete their life cycle. In contrast, there are solitary insects where male and female insect can complete their life cycle without a specific nest or affinity with other individuals. It is a common perception that solitary insects were the ancestors of honeybees and some genetic inquiries supported this idea (Hossam and Shara 2015), while others considered that the origin of bees is from the sphecid wasps (Gullan and Cranston 2010). Bees are totally different from other social insects or even solitary insects, and insufficient studies have been made for studying different aspects of bee's life.

The honeybees are mainly those insects which produce honey as one of their hive products. The different species of honeybees available in India are *Apis mellifera* (Italian bee), *Apis cerana* (Indian bee), *Apis dorsata* (Rock bee), *Apis florea* (dwarf bee) and stingless bees. Besides these bees, bumble bees, *Xylocopa*, Halictid bees are important insects which are the need of today's world agriculture and horticulture.

Honeybees

Honeybees belong to class Insecta, order Hymenoptera, tribe Apini and genus *Apis* having species spread throughout the globe. In India, mainly four

Apis dorsata and *A. labriosa* — *Apis cerana* — *Apis florea* — *Apis mellifera*

Fig 1.1. Different types of honeybees (domesticated and wild).

species of honeybees are available. These are *Apis cerana* (Indian honeybee), *Apis dorsata* (rock bee), *Apis florea* (little bee) and *Apis mellifera* (Italian bee). These all four species bear sting. Besides these honeybee species, stingless bees (do not bear sting) are also available in India which produces honey. Among all these species, *Apis dorsata* and *Apis florea* are wild where they opt for open nesting and cannot be domesticated in modern hives. They construct single vertical combs, whereas *Apis mellifera* and *Apis cerana* can be domesticated in hives and they construct parallel combs in cavities or in modern hives. Stingless bees can also be reared easily under domesticated conditions.

Apis cerana – Indian bee

Distribution: *Apis cerana* is found in North east region of India (Nagaland, Arunachal Pradesh, Manipur, Assam, North Bengal, Tripura, Sikkim, Meghalaya and Mizoram), North India (Himachal Pradesh and Jammu & Kashmir) and South Indian states (Tamil Nadu, Kerala, Andhra Pradesh and Karnataka). *Apis cerana* have three subspecies in India: *A. cerana indica*, *A. cerana cerana* and *Apis cerana himalaya*.

Morphology and honey yield

The nest consists of several parallel combs with a bee space of 7-9 mm. Combs are about 25 mm thick. The brood cells are about 11.5 mm deep and vary in width from 4.17 to 4.83 mm. Tongue length varies from 4.39-5.53 mm. They make 21.25 cells per 10 linear cm comb. The total developmental

Tetragonula iridipennis *Lepidotrigona ventralis*

Lophotrigona canifrons *Tetragonula gressitti*

Fig 1.2. Different types of stingless bees in India.

period for worker is 18-19 days, 24 days for drone and 15-16 days for queen. *Apis cerana* yields about 3-5 kg honey/colony. The flight range is 0.8-2 km. These bees are prone to absconding, swarming, robbing and development of laying workers.

Apis mellifera – Italian bee

Distribution: *Apis mellifera* is not the native bee of India. It was imported from Italy and successfully established at Nagrota Bagwan in India by A S Atwal and Sardar Singh in 1962. Now *A mellifera* is established in most parts of India, viz. Himachal Pradesh, Jammu & Kashmir, Punjab, Haryana, Delhi, Uttar Pradesh, Madhya Pradesh, West Bengal, Odisha, Uttarkhand, Chhattisgarh, Gujarat and few pockets of southern India.

Morphology and honey yield

These bees are yellowish in color having black bands on their abdomen. The nests of *Apis mellifera* are similar like those of *Apis cerana*. They prepare parallel combs of 10 mm bee space. Colonies are large usually between 15000 to 60000 workers. The body weight of worker is 90-120 mg. Tongue length ranges from 5.4-7.2 mm. They carry nectar load of 40-80 mg and pollen load of 12-29 mg. the flight range is 2-5 km. Swarming and absconding instinct is less. The average honey yield is 10-15 kg/colony which can be increased up to 40 kg/colony through migration.

Apis cerana natural nest — Apis cerana comb

Apis mellifera hive — Apis mellifera frame

Fig 1.3. Hive and comb structure of *Apis cerana* and *Apis mellifera.*

Apis dorsata – Rock bee

Distribution: The distribution of *Apis dorsata* is similar to that of little bee. It is found throughout the plains and subtropical parts of the country up to altitudes of 2000 m.

Morphology and honey yield

The giant honeybees are largest of all honeybee species found in India. They prepare single comb of about 1 m diameter hanging on the tree branches and rock cliffs (3-25 m above the ground). These are migratory in nature and travels up to 100-200 km distance every year. The nest population ranges from 60000-100000 workers. The worker bees look like the queen of *Apis mellifera*. The workers are capable of foraging during night time when full moon light is available. *Apis dorsata* produces 60-100 kg of honey in full grown nest and average about 20-50 kg.

Apis florea – Little bee

Distribution: The distribution area of *Apis florea* is generally confined to warm areas.

Fig 1.4. Natural nest architecture and site of *Apis dorsata.*

Morphology and honey yield

The dwarf honeybee is the smallest species among all bees both in body size of worker bees and size of the nest. The workers are considerably smaller than the drones and queen bees although the later castes are almost as big as those of *Apis cerana*. The queen can normally lay 350-700 eggs per day. They have a short flight range of about 100 m from the nest. The dwarf bees are migratory in nature and they can thrive in temperatures up to 50°C. The colonies of *A. florea* produce only a small quantity of honey, i.e. 500-750 g.

T. gresitti – Stingless bees

Stingless bees are also known as dammer bees. These are smaller in size having reduced stringer.

Distribution: Stingless bees are mostly found in tropical parts of the world. In India, these inhabit North Eastern states and South Indian Peninsula. About nine species of stingless bees are reported from India. Out of which six species are reported from Nagaland only (Chauhan and Singh 2021).

Morphology

Stingless bees are very small in size as compared to other honey producing bees. The colony consists of queen, workers and few drones. These are

Apis florea natural nest   *Apis florea* exposed comb

Fig 1.5. *Apis florae* natural nest architecture with exposed comb.

Fig 1.6. Natural nest architecture of stingless bees (*T. gressitti*).

yellowish brown to black in colour. Few species bear coloured bands on their abdomen. These bees make nests in tree cavities and underground also. The natural nest architecture showed nest entrance oriented to south while some species make entrance to North West. The involucrum is made up of wax, cerumen mixed with mud. The brood cells are placed in the centre stacked over one another in horizontal manner. The size of queen cell ranges up to 7.15 mm in length and 4.84 mm width while the worker and male cells are smaller in size (4.16 × 3.14 mm and 4.06 × 3.09 mm). The honey pots (13.31 × 12.91 mm) and pollen pots (14.25 × 13.95 mm) are bigger in size than other cells present in the colony. These bees produce low quantity honey, i.e. 250 g/nest.

Pollen bees: All the different honeybee species are exceptional pollinators in their own capacity besides being honey collectors. In addition to these, there are more than 20000 species of other bees which help in pollination. It is cleared that all bees are not honeybees. Batra (1995) has even separated non-apis bees in a separate group of pollen bees and that group include all bees apart from honeybees which help in pollination.

Bumble bees

Distribution: Bumble bees are found mainly under temperate to subtropical regions of the globe. While, some perennial species of bumble bees are reported in the tropical areas also. In India, approximately 48 species of bumble bees are present in hilly regions of Himachal Pradesh, Jammu

Fig 1.7. Scientific hive of stingless bees.

& Kashmir, Uttarakhand, West Bengal and all North Eastern states of the country (Saini *et al.* 2011; Chauhan 2011, 2015). *Bombus haemorrhoidalis* is the most dominant species of bumble bee in North Himalayan region and most of the rearing experiments are done with this species of bumble bee.

Morphology

Bumble bees are much bigger in size than honeybees and stingless bees. Bumble bees live either underground in natural cavities or over ground in dry vegetation of thick dry leaves and wooden twigs (Fig 1.9). The colony consists of a queen, hundreds of workers and few drones (Fig 1.8). They possess differently coloured banded abdomen and the colour of the bands vary from one species to another. The colony start raising queen cells either in the autumn or in the late summer. The respective length and width of the queen cell were 21.97mm±1.04 and 12.90mm±0.99. The brood cells are found to be oval in shape. The adult queens, at the time of emergence, were found to have white and orange sclerite bands on their body which later turns to yellow and orange bands. The posterior end of their abdomen is blunt and bears sting. The drones are morphologically different from queen and workers because they are smaller in size with yellow or white pubescence on their head (vertex), thorax (lateral sides) and sometimes on their legs also.

Fig 1.8. Different castes of bumble bee, *Bombus haemorrhoidalis*.

They have a third antennal segment that is longer than workers and queen. The posterior end of their abdomen is round in shape. Drones do not sting as they do not bear sting apparatus. Drones leave the nest permanently after 2-3 days of emergence. In appearance, the workers of bumble bee are similar to queens unless for their smaller size than queens. The posterior end of their abdomen is blunt and bears sting. It was observed that workers do all the duties of the nest except for laying of eggs so long the queen is alive. When the queen died or failed at to lay eggs, the workers also started egg laying.

Importance of bumble bees

Bumble bees are found to be efficient and reliable pollinators of crops grown under protected conditions (fully automated) because they can forage

Fig 1.9. Fully grown laboratory reared bumble bee colony.

even under adverse conditions of weather. They have a short flight range and consist of a low population. They are also valuable in buzz pollination of the crops (Corbet *et al.* 1982, 1988; Erikson and Buchmann 1983). It has been proved that the largest quantified value of bumble bees comes from their pollination services to agriculture, which are worth billions of dollars annually (Dias *et al.* 1999; Goulson 2003; Winter *et al.* 2006). These bees are widely used for the pollination of greenhouse crops in Holland, New Zealand, China, Japan, Bulgaria, UK and Germany. In India, several bumble bee species, viz. *B. waltoni, B. haemorrhoidalis, B. kereinsis, B. himalayanus, B. asiaticus, B. personatus, B. rufofaciatus, B. trifaciatus* and *B. tumicatus* are present. But *B. haemorrhoidalis* is a dominantly available species spread everywhere from low-lands to high altitude region of Himalayas (Saini and Ghattor 2007). Many attempts for its domestication have been made since the last one decade (Dayal and Rana 2004; Thakur *et al.* 2005: Chauhan 2011; Chauhan and Thakur 2014, Chauhan *et al.* 2013, 2014, 2015, 2016). The experiments on the rearing of bumble bees (*Bombus haemorrhoidalis* Smith) were started in 2000 and for the first time in the country, bumble bees were reared throughout the year during 2012-2013 (Chauhan *et al.* 2014). The domesticated bumble bee colonies were used extensively for pollination of crops like cucumber (Chauhan 2011; Kashyap 2008), tomato (Yankit *et al.* 2018), kiwi (Nayak 2018) and capsicum (Thakur 2018) under protected conditions.

Xylocopa

Distribution: These bees are also known as carpenter bees due to their nesting behaviour. These bees are distributed mainly in the subtropical and tropical regions of the globe but few species are recorded from the European countries. More than 500 species are present globally and 40 of them are reported from India.

Morphology

Xylocopa exhibits both solitary and social behaviour. The size of the adults is similar to bumble bees but these do not possess bands on their abdomen. Some species have coloured pubescence on dorsal side of thorax. These are active between February to October and in winter the adult queens/males undergo hibernation. The bees are generally dark metallic in colour and polymorphism is exhibited by different species (Fig 1.10). The social *Xylocopa* species have queen, males and few workers which can be distinguished.

Importance

Xylocopa are good pollinators of cucurbit crops and can also be used for the pollination of tomato under protected conditions. In India research trials on pollination are going on in Bihar, Assam and Nagaland to establish the pollination efficiency of this pollinator. In India, *Xylocopa fenestrata*, *X. leucothorax* and *X. tenuiscapa* are under experiments for evaluating pollination efficiency.

X. fenestrata *X. leucothorax* Natural nest (pc. aau)

Fig 1.10. Different species of *Xylocopa* and their natural nest architecture.

Halictid bees – Sweat bees

Distribution: Halictid bees are found worldwide in distribution but are especially abundant in temperate regions of the world. It is one of the largest bee groups consisting of 3600 described species worldwide. Halictidae is divided into four subfamilies: Rophitinae, Nomiinae, Nomioidinae and Halictinae. In Florida, there are 44 known species found within eight genera (Pascarella 2008). In India, total 211 species were reported so far belonging to different genera and subfamilies.

Morphology

Among all the six important bee families under the order Hymenoptera, Halictidae are important ones. These are also called as sweat bees and are a very varied group of metallic and non-metallic bees. They are typically more abundant than most bees except *Apis* (honeybee) species. Halictids display the most diverse gradation in social behavior (Michener 2007) since their different species are solitary, communal, semi-social or eusocial. Some species exhibit solitary or eusocial behavior depending on the time of year, geographic location, altitude and often unknown factors (Michener 2007). The genus *Lasioglossum* (=*Dialictus*) in Halictidae is one of the largest genera of bees worldwide, with an incredibly diverse array of behaviours.

Most of the halictids dwell in underground nests. They also built nests in semi rotten wood that usually resemble ground nests (Eickwort and Eickwort 1973). Some species that are parasites of other bees do not build nests at all. Although they usually nest in the ground, their nesting behavior is highly variable from species to species. Each subfamily has its own characteristic nesting habit:

- In Rophitinae, larval cells are typically excavated at a slant or horizontal aligned to the main lateral tunnel.
- Nomiinae build vertical cells, sometimes in clusters.
- Nomioidinae make nest in sub-horizontal cells in vertical burrows.
- Halictinae have diverse nest types, including series of cells, scattered cells and clusters of cells. The area around the clusters sometimes is excavated. Cells usually are lined with a waxy substance likely extruded from the bee's Dufour's gland (a gland found on the underside of the abdomen). Nest architecture of halictids that build nest in rotting wood is irregular, usually due to the nature of the wood (http://entnemdept.ufl.edu/

Halictus sp. on citrus

Halictus lucidipennis

Halictus sp. on mustard

Sphecodes sp. (PC. Manju)

Fig 1.11. Different species of halictid bees.

creatures/misc/bees/halictid_bees.htm). These Halictids typically nest in barren soil and prefers a sunny location. It may be noted that minimum tillage and insecticide use will help to increase populations of Halictidae and other soil nesting bees.

Importance

Sweat bees are exceptional pollinators for many crops and wildflowers, especially stone fruits, pome fruits, sunflower and alfalfa (Fig 1.11). Sweat bee populations can be increased tremendously through planting wildflowers and offering nesting areas.

Importance of bees

Insects are dominant animals on this earth. The immense benefits of insects in maintaining economy outweigh the injury inflicted by them. Among them, honeybees are one of the very few insects directly beneficial to man. For instance, honey is a highly valued food product from honeybees and it is also used as medicine. In addition to honey, other products like bee's wax, pollen, royal jelly and bee venom are also produced by bees from colonies. However, more than the value of these hive products, bees perform a very important objective of nature by assisting in pollination of plants while foraging for its food on flowers.

Of all the different pollination agents, the insects are mainly responsible for effective pollination and are considered as the most reliable and efficient pollinators. These insects such as honeybees, bumble bees and solitary bees are in commercial use for pollination. Bees are responsible for pollinating plants that provide most of our food and it is found that 30% of food for human consumption originates from plants pollinated by bees (Anonymous 1997). Most of the cereal crops, vegetables (radish, cabbage, green pepper, tomato) and fruits (apple, pear, citrus, kiwi, melons etc.) are dependent upon insect pollinators (McGregor 1976; Free 1993). Honeybees have a dominant role in commercial pollination around the world. Apart from honeybees, wild bees (bumble bees, stingless bees, solitary bees etc.) are also valuable pollinators of crops.

In naturally open conditions, the honeybee pollinators can be managed, but under protected conditions, a number of pollination problems are encountered due to their long-flight range, high population and adverse temperature conditions (Dag and Kammer 2001; Chauhan 2011; Thakur 2011). In India, honeybees mostly are used for the pollination of various agricultural and horticultural crops but in countries like Netherland, China, Turkey, Brazil, Europe and USA, bumble bees are widely used for the pollination of many crops such as tomato (Asada and Ono 1997), capsicum (Shipp *et al.* 1994; Porporato *et al.* 1995; Abak *et al.* 1997; Meisels and Chiasson 1997; Dag and Kammer 2001; Kwon and Saeed 2003; Ercan and Onus 2003), egg-plant (Abak *et al.* 1995), melons (Fisher and Pomeroy 1989), water melon (Ravestijn and Kraemer 1991; Stanghellini *et al.* 1997, 1998a, 1998b, 2002), cucumber (Stanghellini *et al.* 1997, 1998b, 2002), strawberry (Paydas *et al.* 2000a), rasp berry (Willmer *et al.* 1994), cran berry (MacFarlane *et al.* 1994b; MacKenzie 1994), blue berry (Whidden 1996; Stubbs and Drummond 2001), apple (Goodell and Thomson 1997; Thomson and Goodell 2001), kiwi (Pomeroy and Fisher 2002), plum (Calzoni and Speranza 1996) both in open (Cufford 1973) and protected conditions (Delplane 1995).

It is a fact that honeybees are not able to forage on flowers without nectars and flowers with deep corolla. Furthermore, they also remain confined to their hives during cool temperature and cloudy conditions. This results in decreased pollination leading to decreased productivity due to lesser activity by honeybees. On the other hand, bumble bees and stingless bees are found to be the most efficient and reliable pollinators for protected crops as they work well under adverse conditions of weather, have short-flight range and consist of low populations (Corbet *et al.* 1982, 1988 and Erikson and Buchmann 1983). Hence, among wild pollinators, bumble bees and stingless bees are emerging as important pollinators. The successful rearing of such pollinators is being done in some countries like Japan, China, Israel, Turkey etc. As of now, only five species of bumble bees, viz. *Bombus terrestris, B. impatiens, B. occidentalis, B. lucorum* and *B. ignitus* are commercially reared worldwide. In

North America, only two species namely *B. impatiens* and *B. occidentalis* are reared and used in pollination (Velthuis and van Doorn 2006). Likewise, stingless bees are also widely used for pollination of crops under protected conditions (Chauhan *et al.* 2019).

Bumble bees and stingless bees have some contrasting characters which make them important pollinators under different climatic conditions. Bumble bees are mostly found in temperate regions so are very well acclimatized to work in colder weather conditions and fully automated greenhouses. On the other hand, stingless bees are tropical bees capable of tolerating high temperatures so more suitable in naturally ventilated greenhouses which are very common in our country.

Above all, the increased use of bumble bees in commercial protected farming with environment controlled greenhouses has further designated it as economically important pollinator for high economic return and production of off-season crops. However, indigenous bumble bee species are considered more important for domestication and pollination services instead of exotic bumble bees in order to avoid possible abiotic and biotic problems. In India, several native bumble bee species, viz. *B. waltoni, B. haemorrhoidalis, B. kereinsis, B. himalayanus, B. asiaticus, B. personatus, B. rufofaciatus, B. trifaciatus* and *B. tumicatus* are present. But *B. haemorrhoidalis* is widely available species spread from lowlands to high altitude region of Himalayas (Saini and Ghattor 2007). Similarly, indigenous stingless-bee species, *Tetragonula iridipennis* and *Lepidotrigona* spp. are most suitable for pollination of crops under protected conditions. Researchers have observed many folds increase in fruit yield when pollinated by stingless bees over honeybee pollination (Chauhan *et al.* 2019). In Australia farmers are utilizing indigenous *T. carbonaria* for pollination of Macadamia and strawberry.

CHAPTER 2

Stingless Bee Diversity

The stingless bees belong to the family Apidae and are close relatives of true honeybees, carpenter bees, orchid bees and bumble bees. Among all bees, stingless bee is the smallest (4.0 to 5.0 mm long) of the honeybees. In contrary to other Apis honeybees, that sting in stingless bees is atrophied in size and does not sting but they have defensive mechanisms to bite when their nest is disturbed by enemies. Stingless bees are known to have the oldest evolved bees found preserved inside ambers which were 80 million years old. Stingless bees are a large group of bees (approximately 600 species) that constitute the tribe Meliponinae. The estimated numbers of species identified thus far included 50 species in Africa, 300 species in the Americas, 60 species in Asia, 10 species in Australia and 4 species in Madagascar (Bradbear 2009).

Stingless bees are reported to be found throughout the world but confined to tropical and subtropical regions such as Australia, Africa, Southeast Asia, and tropical America. These bees are active all the year round, but are less active in cooler weather regions. According to Roubik (1989), local and regional diversities are much high in the Neotropics where it is reported that up to 60 meliponine species can be found in a single forest.

Stingless bee fauna

Stingless bees constitute a large group of bees with around 500 described species (Michener 2013), comprising the tribe Meliponini and sub-tribe Meliponina. These belong to the family Apidae and are close relatives of common honeybees and bumble bees. Meliponines have stingers but are significantly reduced and cannot be used for self-defence purpose. Stingless bees have many genera. Some of their genera can be found in tropical or subtropical regions of the world (Silveira *et al.* 2002). The workers which are the females, possess feeble or vestigial stingers but are unable to cause pain with their stingers. Hence, the term "stingless" is being used to entitle the species. Some species have adequately stronger mandibles to cause a mild bite or even crawl into ears and nostrils of the trespassers while others release a caustic liquid from the mouth that can cause intense irritation when in contact with the skin (Rahman *et al.* 2015).

Rasmussen (2013) studied the distribution of stingless bees throughout the Indian subcontinent and concluded that stingless bees are present in many regions of the Indian subcontinent, except at higher altitudes or the drier interior regions. He found nine species of stingless bees in Indian subcontinent: *Lepidotrigona arcifera* (Cockerell), *Lisotrigona cacciae* (Nurse),

Lisotrigona mohandasi Jobiraj and Narendran, *Tetragonula* aff. *laeviceps* (Smith), *Tetragonula bengalensis* (Cameron), *Tetragonula gressitti* (Sakagami), *Tetragonula iridipennis* (Smith), *Tetragonula praeterita* (Walker) and *Tetragonula ruficornis* (Smith).

Stingless bees are extremely developed social insects and live in a colony under organized system with appropriate division of labour. Some species have groups of as many as 80,000 individuals and other less than 100 (Rahman *et al.* 2015). The majority of stingless bee species do not disturb man and they can be manipulated safely for domestication in the homestead garden (Rahman *et al.* 2015). All the different species of stingless bees produce honey at an average of around 600-700 g/year (Kumar *et al.* 2012) and 200-500 g per season by each colony (Rahman *et al.* 2015), having high quality medicinal value. It is proven that stingless bees are effective pollinators of tomato, pepper, strawberry, citrus fruits, melons and other harvests and adapted well to protected conditions in different regions around the globe (Maeta *et al.* 1992; Sarto *et al.* 2005; Cruz *et al.* 2005; Antunes *et al.* 2007; Santos *et al.* 2009; Giannini *et al.* 2015b; Chauhan and Singh 2021). Silveira *et al.* (2002) observed that the flight activity of stingless bee started at relative humidity 80-89% and temperature 17°C and 22°C. The minimum temperature requirement for exit of the bee from hive is 11°C. The stingless bees mostly nests in cavities, old walls, inside the termite mounds, tree trunk and underground cavities (Nogueira-Neto *et al.* 1997; Eltz 2003; Roubik 2006). Worldwide, the total number of stingless bee species described is about 500 species (Michener 2013).

Stingless bees of Brazil

Brazil is home to hundreds of stingless bee species that belong to Meliponini, with around 300 species already identified and more are still undiscovered. A total of 244 valid species, and about 89 undescribed affiliated to 29 genera, are recorded from Brazil (Pedro 2014). These stingless bees vary greatly in size, habits and shape, and among them 20 to 30 have good potential as honey producers (Pedro 2014). Among many others, found throughout the country, species such as *Melipona subnitida* and *Melipona scutellaris* in the northeast region of the country, *Melipona quadrifasciata* and *Melipona rufiventris* in the south-southeast, *Melipona compressipes manaosensis* and *Scaptotrigona polistycta* in the north and *Tetragonisca angustula* are increasingly kept by small, medium, and large producers.

Stingless bees of Australia

In Australia, most common group of bees in north tropical and sub-tropical Australia are the stingless bees (Heard and Dollin 2000). These bees belong to two genera *Trigona* and *Austroplebeia* (Michener 1990). Altogether, six species under *Trigona* (Heterotrigona) occur in Australia (Dollin *et al.* 1997). The *Austroplebeia* is found only in Australia and they consist of about four species. The Australian stingless bee species are also small, measuring less than 4mm in length, black in colour and nest in hollow of trees. The most common species found in Australia are described to be *Trigona carbonaria* (69%), *T. hockingsi* (20%) and 11% for *T. mellipes* and *T. clypearis*. The different stingless species appear quite similar to each other except *Trigona carbonaria* and *Austroplebeia australis*. They display the utmost variation, as the latter is smaller and less active. Among them, *T. carbonaria* is the most widely studied bee for pollination research.

Stingless bees of Africa

All African stingless bees (Meliponini) are social in nature. There are six genera (Eardley 2004) of African stingless bees that comprises 19 species namely, *Cleptotrigona, Dactylurina* Cockerell, *Meliponula* Cockerell, *Plebeina* Moure, *Hypotrigona* Cockerell and *Liotrigona moure*. In five of the genera (*Dactylurina* Cockerell, *Meliponula* Cockerell, *Plebeina* moure, *Hypotrigona* Cockerell and *Liotrigona* moure), worker bees collect pollen and nectar from flowers but in one genus (*Cleptotrigona* moure) their peculiarity is that they rob pollen and nectar from the nests of other stingless bees. The different stingless bee species under domestication in Kenya include *Meliponula ferruginea, Meliponula bocandei, Meliponula lendliana, Plebeina hildebrandti* and *Dactyrulina schmidti* (Eardley 2004). Domestication has been envisaged for honey production and as an incentive towards forest conservation (Kiatoko *et al.* 2014; Kiatoko *et al.* 2016).

Stingless bees in Indian subcontinent

In India, the different stingless bees belong to the order Hymenoptera under the family Apidae, sub-family Apinae and Tribe Meliponini which has two main genera, viz. *Melipona* and *Trigona*. Altogether 250 species have been identified throughout the Neotropical and Indo-Burma-Malayan region. These stingless bees are distinguished from corbiculate Apinae by combination of reduced fore-wing venation and the presence of a jugal lobe in the hind wing. The present state of knowledge on stingless bees of India, their diversity and foraging plants are not clearly known. Bingham (1897) described most of the Indian species under Melipona almost a century ago. There is sporadic information on stingless bees and the earliest accounts were given by Sakagami (1978) and Rasmussen (2013) for continental Asia and Indian subcontinent. Out of the stingless bee species in India, only the two genus, Trigona Jurine and Lisotrigona Moure have been found discovered in the Indian subcontinent (Michener 2007). According to Rasmussen (2013), all together eight named species of stingless bees are known from the Indian subcontinent, viz. *Lepidotrigona arcifera* (Cockerell), *Lisotrigona cacciae* (Nurse), *Lisotrigona mohandasi* Jobiraj and Narendran, *Tetragonula laeviceps* (Smith), *Tetragonula bengalensis* (Cameron), *Tetragonula gressitti* (Sakagami), *Tetragonula iridipennis* (Smith), *Tetragonula praeterita* (Walker), and *Tetragonula ruficornis* (Smith) in Indian subcontinent.

The distribution pattern of stingless bees in India was studied by Rahman *et al.* (2015). They have reported four species of stingless bees prevalent in South India namely *Tetragonula. iridipennis, T. laeviceps* and *Lepidotrigona arcifera* while *Tetragonula iridipennis* and *T. laeviceps* were present in the North-west India. The study also revealed that *Tetragonula iridipennis* and *T. laeviceps* were most commonly available species in all selected zones of India. The nesting behaviour of *Tetragonula irridipennis* was usually found to nest inside timber or hollow log cavities, whereas, *T. bengalensis* constructed nest in bamboo and others make in cracks or crevices of stone and mud walls in India. Most of the researchers concentrate purely on stingless bee biology, morphometry, natural enemies and its pollination biology (Muthuraman and Saravanan 2004; Danareddi and Viraktamath 2009; Kumar *et al.* 2013), or species diversity in India (Rahman *et al.* 2015). From Kerala, two stingless bees are mainly reported, viz. *Tetragonula irridipennis* by Raakhee and Devanesan (2000) and *Lisotrigona mohandasi* by Jobiraj and Narendran (2004).

Stingless bees of North-east India

The North-eastern Region of India wich is enclosed within a diverse natural habitat with varied topography, climate and forest is well known as a part of the Indo-Myanmar biodiversity hotspots (Kumar 2014). In this region, Nagaland and Arunachal Pradesh are the hill states of North-eastern India where the studies on stingless bees in respect of their domestication and use in pollination under protected condition in vegetable production are highly essential as of today.

In North-east India, five species were reported by Rahman *et al.* (2015), viz. *Tetragonula bengalensis, T. iridipennis, T. ruficornis, T. laeviceps* and *Lepidotrigona arcifera,* whereas *Lepidotrigona arcifera* Cockrell is recorded from Arunachal Pradesh.

In Nagaland, Singh (2016) has observed presence of three species namely *Tetragonula iridipennis, T. laeviceps* and *Lophotrigona canifrons.* However, later

Table 2.1. Distribution of stingless bee species in different regions of Nagaland, India

Species	Approxmiate elevation range (amsl)	Coordinates	Areas
Tetragonula iridipennis	110-1464 m	25° 39'10.398" N 25° 42'53.927" N 94° 1'22.812"E 94° 2'25.525"E	Sovima, Ingtanki, Choumukedima, Gaili, Merapani, Naginimora, Namsa, Suring, Tuli, Aonokpu, Punglawa, Baghty, Bhandari, Zubza, Changtongya, Chare, Doyang area, Kiphire, Jalukie, Mangkolemba, Medziphema, Meluri, Mon, Pangsa, Tobu, Tsurangkong, Tizit, Akuloto, Changki Range, Chen, Chessore, Mima, Khonoma, Chozuba, Longleng, Tamlu, Bhumnyu, Longwa, Mokokchung, Peren, Sanis, Sangsang, Tuensang, Tseminyu, Wakehing, Wahezo, Wokha

(Continued)

Species	Approxmiate elevation range (amsl)	Coordinates	Areas
Lepidotrigona ventralis/ L. arcifera	458-2018 m	25º 39.894 N 25º 34'15" N 93º 51 .268 E 94º 17'43" E	Tuzatsu, Aonokpu, Punglawa, Baghty, Bhandari, Zubza, Changtongya, Chare, Doyang area, Kiphire, Jalukie, Mangkolemba, Medziphema, Meluri, Mon, Pangsa, Tobu, Tsurangkong, Tizit, Akuloto, Changki Range, Chen, Chessore, Mima, Khonoma, Chozuba, Longleng, Tamlu, Bhumnyu, Longwa, Mokokchung, Peren, Sanis, Sangsang, Tuensang, Tseminyu, Wakehing, Wahezo, Wokha, Aghunato, Chentang, Khonoma, Chenmoho, Kidima, Helipong,Japhu range, Kikruma, Kohima, Longkhim, Noklak, Pfutsero, Phek, Saramati, Surohoto, Shammatore, Tokiye, Zunheboto
Lophotrigona canifrons	177- 910 m	25º 38.799 N 25º 42'28.752 93º 51 .268 E 93º 33'37 .16" E	Aonokpu, Punglawa, Baghty, Bhandari, Zubza, Changtongya, Chare, Doyang area, Kiphire, Jalukie, Mangkolemba, Medziphema, Meluri, Apio, Mon, Pangsa, Tobu, Tsurangkong, Tizit
Tetragonula laeviceps	662- 830 m	25º 38.827 N 25º 34'15" N 93º 51 .268 E 94º 17'43" E	Pungalwa, Apao, Medziphema village, Jalukie, Pherema
Tetragonula gressitti	956- 1154 m	25º 37'49" N 93º 32'40" E	Kizari, Tuzatsu, Phek

Source: Chauhan (2021).

on these were updated to six stingless bee species, viz. *Tetragonula iridipennis, L. ventralis, T. laeviceps, L. arcifera and Lophotrigona canifrons* and a recently included species, *T. gressitti* Sakagami was recorded (Chauhan and Singh 2021).

Stingless bees are present in subtropical to tropical areas of the planet and distributed from lowlands to high hills. The distribution range of stingless bees on the basis of altitude has shown a gradient that extends from 300 to 2000 m, covering the dry tropical forest in the warmest section and mixed pine-oak forest in the temperate extreme of the range. The species limited to the warm zone (300–1600 m) are *Scaptotrigona hellwegeri, Trigonisca pipioli, Frieseomelitta nigra.* The species limited to the temperate zone (1700–2000m) are *Partamona bilineata, Plebeia fulvopilosa, Nannotrigona perilampiodes, Melipona fasciata.* Lastly, there are two species broadly distributed (300–2000 m), viz. *Geotrigona acapulconis* and *Lestrimelitta chamelensis* (Gonzalez *et al.* 2016). In India, stingless bee mapping reveals the altitudinal variations at species level. A total of six species were recorded while observing for diversity of stingless bees in Nagaland. These are *Tetragonula iridipennis*, *Tetragonula laeviceps*, *Lophotrigona canifrons*, *Lepidotrigona ventralis*, *Lepidotrigona arcifera* and *Tetragonula gressitti*. The distribution of these species ranged between 110 m-2018 m above mean sea level (Table 2.1) (Chauhan and Singh 2020). Out of these species, *Tetragonula iridipennis* is present in lower altitudes while *Lepidotrigona ventralis* is present in higher altitudes.

■

CHAPTER 3

Nest Architecture and Identification of Stingless Bees

Stingless bees are highly social insects living in colonies. Beekeeping with stingless bees is called Meliponiculture, which has been practised for many centuries in various parts of the world. Although least explored, stingless bees have also been kept for centuries in India, Sri Lanka and Nepal (Crane 1999; Kumar *et al.* 2012). The widespread stingless bee species in the Indian subcontinent *Trigona iridipennis* is used for meliponiculture where colonies are kept in tree logs, wooden boxes, and clay pots for harvesting small quantities of highly prized medicinal honey and also for the wax and propolis, produced and gathered by the bees besides being used for household and curative properties (Crane 1999). Honey production in stingless bees' species is specific as they have their unique characteristics which is highly valued as medicinal honey compared to that of other honeybees.

Stingless bees are perennial in nature and a colony is usually consists of hundreds to thousands of workers. The proper presence of nesting sites is essential for survival, maintenance and reproduction of stingless bees (Hubbell and Johnson 1977; Batista *et al.* 2003; Eltz *et al.* 2003; Roubik 2006). Although their nesting habits varied enormously, most of the stingless bee species build their nests in several substrates, such as subterranean cavities, tree trunks, rock crevices, brick walls, abandoned termite nests, arboreal ant nests and subterranean chambers abandoned by ants (Campos 1987; Roubik 2006). A few species build their nests in exposed positions (Schwarz 1948; Michener 2013). The main material for construction of nest is cerumen, a mixture of bee wax and plant resins while they also use bitumen, a mixture of mud, plant resins, animal faeces etc. (Rasmussen 2013). The nest entrance of stingless bees varied in shape, length, and colour. For example, nest entrance of *Heterotrigona itama* is funnel shape, whereas it is round mount-like shape in *Geniotrigona thoracica* (Syafrizal and Yousuf 2014; Kelly *et al.* 2014). The common structure of stingless bee nests consists of entrance tunnels, brood cells, food storages (honey and pollen cells), cerumen and bitumen layers (Sakagami *et al.* 1983; Michener 2007). *Melipona* and other genera have narrow nest entrances which allows the nest to be defended by one or only a few guards positioned at the mouth of the entrance tube and variation of nest entrances are related to defence and foraging activities of stingless bees (Biesmeijer *et al.* 2005).

Location and collection of wild stingless bee colonies

Different researchers described different methods of location and collection of wild stingless bee colonies. The whole colony of stingless bee could be located and collected from their natural hives and relocated into meliponiaries or colonies are transferred into artificial hives. Naga beekeepers usually searched the presence of flowering plants and if bees are seen foraging, they would search the colony at the vicinity of foraging area preferably in 100 m. Once hives are spotted, they would fell the tree and cut out the portion of hives by hand-saw and transported them to apiary in the evening. In Tuensang district where *Lepidotrigona arcifera* is found, the entrance tubes were found at a level of 1-6 m above the ground and were slightly slanting downward in nature. The nests were found in tree cavities and the colour of the entrance tube varied from dull white to yellowish brown. The apex of the entrance was wide, more or less round and resembled a trumpet and the average length of the tubes was 5.8 ± 2.0 cm in all the colonies. The tube aperture was funnel-shaped, soft, thin-walled with a smooth surface (Kumar 2014).

Height preference for nesting

Stingless bees have a preference for selection of nesting sites at particular heights. The colonies of *T. iridipennis* make nests at a mean height of 2.57 feet above the ground while *L. ventralis* were nesting at an average height of 5.60 feet (Chauhan *et al.* 2019).

Nest architecture

External entrance: The type and shape of entrance of stingless bees is the important identifying feature as different species make unique type of their entrances. The entrance funnel of *Tetragonula iridipennis* is dark brown to black in colour with an opening diameter of 1.81 ± 0.14 cm while *L. ventralis* funnel is white in colour with an opening diameter of 3.52 cm and length ranging from 4.16 cm – 22.00 cm with mean length of 11.65 ± 2.54 cm. The entrance funnel is made up of resins and wax in *T. iridipennis* while in *L. ventralis* it is constructed with wax only. Average 9.66 guard bees are recruited at the entrance funnel in *T. iridipennis,* whereas in case of *L. ventralis*, 6.79 workers are recruited.

Activity of foragers at nest entrance

Foraging activity of stingless bee, *T. iridipennis* at the nest entrance is maximum in late morning hours (1000 h) while minimum at early morning time. Bee activity of *L. ventralis* recorded maximum in morning hours while minimum at late noon hours (1400 h). The incoming activity was maximum at evening hours and outgoing was maximum during late morning hours of the day generally in all the stingless bee species.

Internal colony structure

According to Chauhan and Singh (2019), the nest entrance in most of the stingless bee species is normally oriented to south. In *T. iridipennis* the

Nest entrance of *Lepidotrigona ventralis*

Nest entrance of *Tetragonula iridipennis*

Nest entrance of *Lophotrigona canifrons*

Nest entrance of *Tetragonula gressitti*

Nest entrance of *Tetragonula* sp.

Nest entrance of *Lepidotrigona arcifera*

Fig 3.1. Diversity in nest entrance of different species of stingless bees.

involucrum is incomplete and made up of wax, cerumen mixed with mud. The brood cells are located in the centre stacked over one another in horizontal manner. The size of *T. iridipennis* queen cell is 7.15 mm length and 4.84 mm width while the worker and male cells looks smaller in size (4.16 × 3.14 mm and 4.06 × 3.09 mm). The honey pots (13.31 × 12.91 mm) and pollen pots (14.25 × 13.95 mm) bigger in size than other cells present in the colony (Tables 3.1 and 3.2). Similarly, other species, *L. ventralis* have queen cells (7.96 × 5.20 mm) bigger than the worker (5.44 × 4.05 mm) and male cells

(5.04 × 4.58 mm). The honey pots (23.15 × 19.09 mm) and pollen pots (24.66 × 21.85 mm) usually present in rear side of the brood. The whole colony (brood cells, honey and pollen pots) covers inside the involucrum made up of cerumen and wax and sealed with resins. Inside this covering, researchers observed two holes that served as entrance and exit for the bees supported by resin and wax pillars. In front part of the fully covered brood, pollen pots (49 ± 9.60) were present while the honey pots (total 174 ± 24.12) present in the rear side connected with tunnels to the brood. The pollen weight per pot found to be 2.67 ± 0.68 g while 2.14 g of honey found available in each sealed wax cell. The brood normally covered fully with an involucrum made of wax and cerumen. It remain arranged horizontally (23 ± 2.34 layers) in layers one above the other. The different cells present in *L. ventralis* colonies were bigger in size as compared to *T. iridipennis* colonies. Barbosa *et al.* (2013) in *G. subterranean* found that the nest involucrum consisted of many layers covering the brood having mean area of 11.79 × 9.91 cm. The brood layers generally supported by pillars. The honey pots (5.27 × 1.50 mm) and pollen pots (5.26 × 1.45 mm) locate themselves around the brood. The honey and pollen volume in closed pots were found to be 6.82 ml and 7.04 g which is less than the total honey and pollen volume found in *T. iridipennis* and *L. ventralis*. The total population was observed to be in the range of 2726-11074 bees/nest in *G. subterranea*.

The sizes of different castes, i.e. queen, males and workers are different in both the species. The queen have yellowish brown swollen abdomen and wings do not cover it fully, however in workers and males, the wings covers the whole abdomen in *T. iridipennis*. In *L. ventralis*, the bees are dark brown to blackish in colour. The length of queen (7.46 mm), male (4.02 mm) and worker bees (4.10 mm) is less in *T. iridipennis* as compared to queen (8.16 mm), male (5.21 mm) and worker bees (5.59 mm) length in *L. ventralis*, respectively.

Identification of stingless bees

The identification of stingless bees is an important step before conducting any research experiments as it eases the procedure require to follow and make

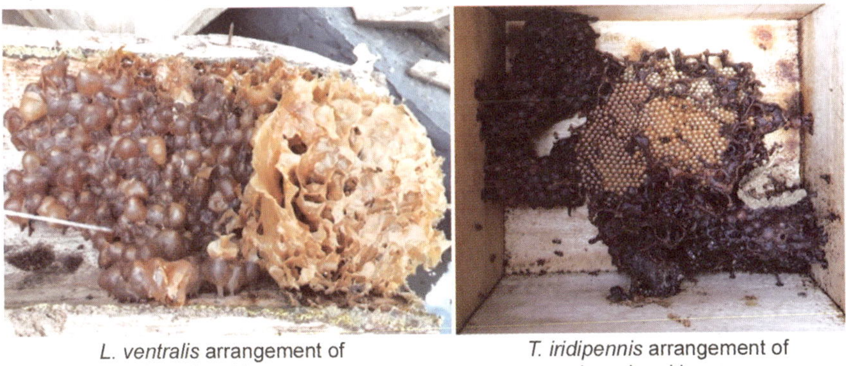

L. ventralis arrangement of brood and honey — *T. iridipennis* arrangement of brood and honey

Fig 3.2. Nest architecture of different species of stingless bees.

Table 3.1. Comparative dimensions of different cells in the nests of stingless bees, *T. iridipennis* and *L. ventralis*

Types of cells	*T. iridipennis*		*L.ventralis*	
	Length ± SE (mm)	Width ± SE (mm)	Length ± SE (mm)	Width ± SE (mm)
Queen cells	7.15 ± 0.21	4.84 ± 0.27	7.96 ± 0.49	5.20 ± 0.17
Worker cells	4.16 ± 0.19	3.14 ± 0.17	5.44 ± 0.66	4.15 ± 0.73
Male cells	4.06 ± 0.11	3.09 ± 0.08	5.04 ± 0.14	4.58 ± 0.21
Honey pots	13.31 ± 0.25	12.91 ± 0.23	23.15 ± 0.48	19.09 ± 0.63
Pollen pots	14.25 ± 0.69	13.95 ± 0.77	24.66 ± 0.22	21.85 ± 0.41

Table 3.2. Comparative dimensions of different castes of stingless bees, *T. iridipennis* and *L. ventralis*

Caste	*T. iridipennis*	*L.ventralis*
	Length ± SE (mm)	Length ± SE (mm)
Queen	7.46 ± 0.09	8.16 ± 0.10
Worker	4.10 ± 0.12	5.59 ± 0.19
Male	4.02 ± 0.14	5.21 ± 0.06

more understanding of the behaviour. Like other insect species, reliable identification of stingless bee species is also possible based on morphological characters. Winston and Michener (1977) studied the morphological characters, evolutionary history and the social behaviour of stingless bees and honeybees. Moure (1961) described the stingless bees into the tribes Meliponini and Trigonini. Characters differentiating the tribes Trigonini and Meliponini based on their body size, usually small, 2 to 8 mm in length and slender (Trigonini) and usually Meliponi, rather-large (from 8 to 15 mm in length) and robust (Wille 1979). Sometimes due to the overlapping of morphological characters in between different species of the same genus, reliable identification of insect species is daunting task and, in such cases, even taxonomist also failed to identify the species correctly (Behere *et al.* 2007). However, due to the advancement of science various molecular techniques have been used to compare the DNA of such taxonomically difficult species. Using DNA barcoding technology, the reliable identification of the taxonomically species is also possible (Hebert *et al.* 2004). DNA barcoding uses the partial fragment of cytochrome oxidase I gene in mitochondrial genome (Hebert *et al.* 2004).

The different methods involved in the identification of stingless bees are:

1. Morphological and morphometrical methods using taxonomic keys.
2. DNA barcoding.

Identification of stingless bee species by morphological features and biosystematics keys

In the Indian region, the few genera of stingless bees can easily be separated using published keys (Sakagami *et al.* 1990; Michener 2007). However, the species-level identification is not trivial. A complex genus of stingless bees is *Tetragonula*, with more than 30 described species and no global identification key (Rasmussen 2008). Specimens of *Tetragonula* from the Indian subcontinent were treated as a single species in the identification key by Sakagami (1978). Older keys, such as those of Bingham (1897) and Schwarz (1939), cannot be used reliably for species identification. Lastly, sometimes external nest features such as nest entrances are the most diagnostic field traits observed, distinguishing closely related species (Camargo and Pedro 2003, 2004; Rasmussen 2004). Kumar (2014) dissected the feral nests of *Lepidotrigona arcifera*. Various key morphological characters were analyzed and identified using adult bees based on previous literature. Within the subgenus *Lepidotrigona,* the "*ventralis*" species group varied based on the smaller body size from other species in the subgenus. The identification of stingless bees can be established with the sight of nest entrance, nest architecture and nesting site (Roubik 2006). For example, very similar morphology was observed in *Ductylurina schmidti* and the two *Meliponula ferruginea* morphs, but their nesting behaviour is distinct. *Dactylurina schmidti* constructs external nest on tree branches while *M. ferruginea* build reddish brown nests in mud walls and trees and *M. ferruginea* have black nests in trees (Eardley 2004; Nkoba *et al.* 2012). That apart, DNA barcoding revealed cryptic genetic variation within the two morphs.

Description of traits for stingless bee identification

Nayak and Prakash (2017) identified the stingless bee, *Tetragonula* spp. of India at the species level using the morphological characters described by Sakagami (1978) and Rasmussen (2013).

Rasmussen (2013) reported that *Lepidotrigona arcifera* which is available in the eastern parts of India, can be identified through diagnostic dense tessellation on head and thorax and densely plumose hairs on the margin of mesoscutum. The worker bee of *Lepidotrigona arcifera* usually have a black spot on each side of the other-wise pale tergite of abdomen and the mesonotum usually enclosed by a border of short thick scale like or tomentose yellowish to whitish hairs. Hair fringes of hind tibia as well as hairs on mid and hind tibiae and basitarsi were brown to blackish. The hairs on posterior margin are very thick and short and simple (Kumar 2014). In the case of *Lepidotrigona ventralis* (Smith), the species can be differentiated by the two lateral and separate dark spots on the otherwise pale metasomal tergum 1 while in *Lepidotrigona flavibasis*, the species can be differentiated by the semi-circle of black integument, partly enclosing the basal depression on the otherwise pale tergum 1. It can also be recognized by the apical metasomal terga brownish to blackish and the fore and middle tibiae with blackish hairs on the external surface (Schwarz 1939).

The *Tetragonula* of the "*iridipennis*" species group is present in all over India and are characterized by having a dark mesoscutum with four distinct hair bands separated by broad glabrous interspaces, and by their smaller

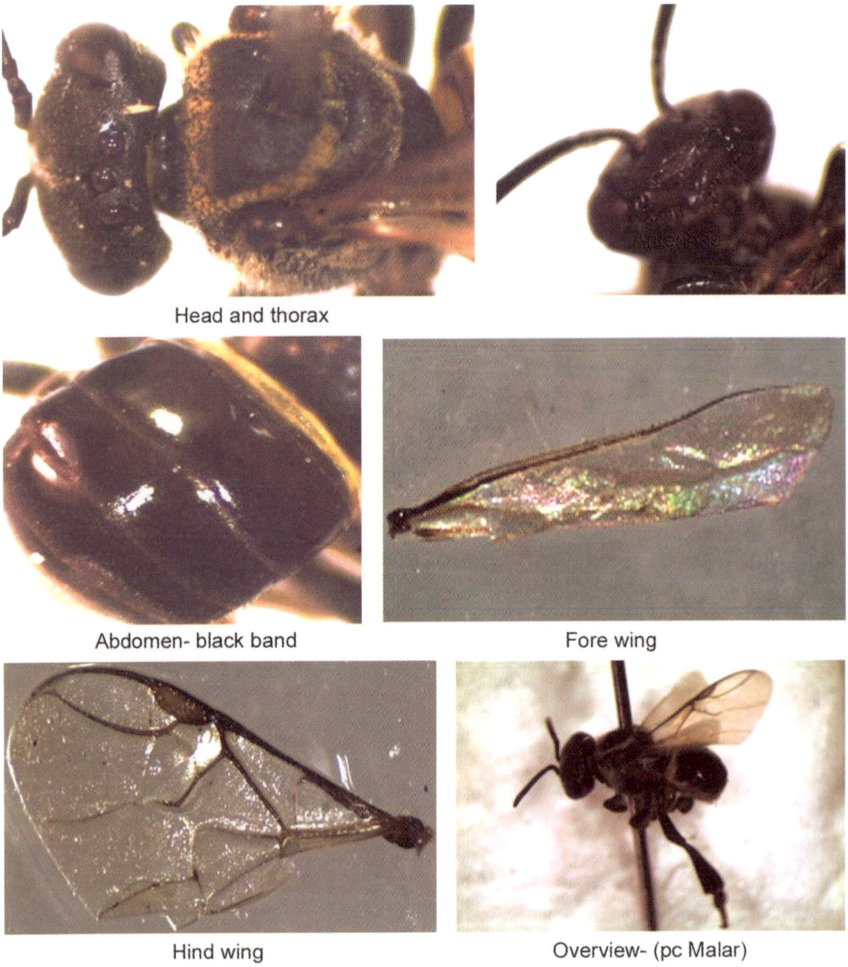

Fig 3.3. Different morphometric characters of stingless bee, *Lepidotrigona arcifera.*

body size, with the forewing length, including tegula, measuring 3.0 to 4.3 mm. Members of *Tetragonula aff. laeviceps* (Smith 1857) usually have a mesoscutum almost as evenly banded with hair as the "*iridipennis*" group, but these species are larger, with forewing length, including tegula, measuring between 4.2 and 4.8 mm. Sakagami (1978) described this species group as the taxonomically most difficult group within *Tetragonula*. Rahman *et al.* (2015) reported that *Tetragonula iridipennis* had blackish head and mesosoma, brownish metasoma; brownish antenna, brownish black legs and their head devoid of pubescence, mesosomal pleuron with brownish short keirotrichia. *Tetragonula bengalensis* is entirely black, with the head black, Mesosoma black and metasoma jet black. Pubescence is generally whitish brown; head with brownish pubescence; brownish antennae; legs are black with dark brown pubescence, tarsus with brownish keriotrichia.

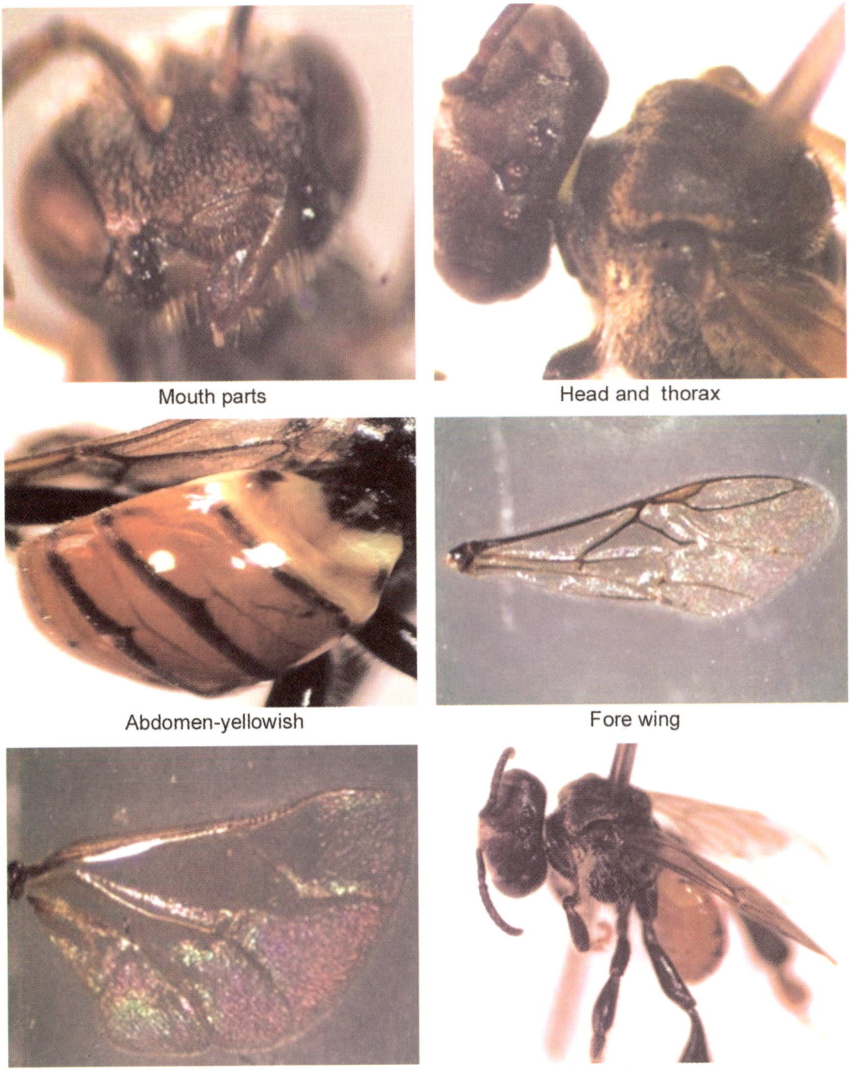

Mouth parts | Head and thorax

Abdomen-yellowish | Fore wing

Hind wing | Overview - *Tetragonula* sp.

Fig 3.4. Different morphometric characters of stingless bee, *Tetragonula iridipennis.*

The worker bees of *Tetragonula gressitti* Sakagami were black coloured with body clearly divided into head, thorax and abdomen as per Chauhan and Singh (2020) (Fig. 3.5). The total body length varied between 6.10-6.11 mm with the head width range between 1.89-1.91 mm. The length of head was 1.65 mm. The compound eyes were big with length of 1.31-1.33 mm and width of 0.47-0.48 mm with the maximum inter-orbital distance 1.28-1.29 mm while the lower and upper inter-orbital distance was 1.00- 1.01 mm and 1.09-1.11 mm. The diameter of median ocellus was 0.21-0.22 mm. The length of clypeus was found as 0.40-0.41 mm and width ranged between

Fig 3.5. Different morphometric parameters of *Tetragonula gressitti.*

0.68-0.69 mm. The oculo-malar space was 0.11 mm. The total length of antennae was 3.09 mm with length and diameter of scape was 0.76-0.78 mm and 0.10-0.11 mm respectively.

The flagellomeres were black and 10 in numbers with 0.88-0.90 mm width. The distance between M-Cu and basal tip of marginal cell was ranged between 1.32-1.33 mm. The length of fore wing ranged between 4.80-4.89 mm however, the hind wing was smaller with length of 3.96-4.01 mm and having seven hamuli with arrangement like 4 + 1 + 2. In hind leg, the length of femur and tibia ranged between 1.04- 1.07mm and 1.96 -1.98 mm while the tarsus was 0.66-0.69 mm. The pretarsus was having 4 tarsomeres with 0.89-0.96 mm in length. The tibia was bearing more black hairs as compared to femur and tarsus. The last tarsomere terminated into claws. The abdomen was black in colour with inconspicuous divisions.

The general colouration of *Tetragonula laeviceps* is black with brownish metasoma, brownish wings, hyaline beyond stigma, deep brownish antenna in base and lighter in the apex. In legs, trochanter was brownish black, tibia and femur were black, brownish metatarsus in black base and brownish tarsomere. The head was with sparse hairs, mesosomal hairs in the pleurite; metasoma was with very sparsely distributed hairs. In legs, the trochanter was with long bristle-like hairs, tibial spur present, femur without spur, metatarsus with rows of keriotrichia and tarsus with keriotrichia.

In *Tetragonula ruficornis,* the head was brownish in colour, mesosoma was brownish black and black metasoma. In the legs, trochanter, femur and tibia were deep brown, tarsus was brownish at the base and lighter in apex. The head was with hairs on the cervix, mesosoma with pleural spur and metasoma without hairs. On the legs, the trochanter was without spur, tibia with bristle-like keriotrichia, femur without spur, tarsus with rows of keriotrichia, metatarsus with sparsely distributed keriotrichia. The legs were testaceous; penicillum testaceous.

Lophotrigona is the underground stingless bees and is till now reported only from Nagaland in India. The species present is *L. canifrons*. They are black in colour with conspicuous fore wing and hind wing. They usually live in environment where moisture content is around 65-70%.

Identification through molecular analysis

Identification of stingless bees is a great challenge as some species are not having characters that are distinctive of a particular species. Ndungu *et al.* (2017) used combined morphological features, morphometrics and

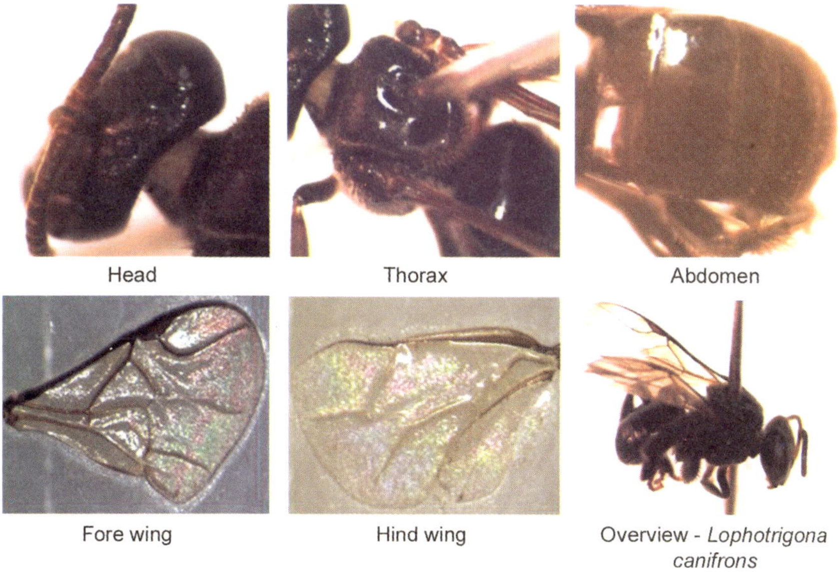

Fig 3.6. Different morphometric characters of underground stingless bee, *Lophotrigona canifrons.*

molecular methods in the identification of stingless bees. They reported that morphometrics alone cannot identify all the stingless bees when identity of the species is completely unknown. By morphometry, they could not distinguish reddish brown *M. ferruginea* and black *M. ferruginea* collected from two different locations as the two morphs were similar in size. Due to the extreme morphological similarities within or between species of stingless bees, the identification of stingless bee species when based on either external or internal morphological characters is a daunting task. However, molecular techniques have provided some additional help to the taxonomist. Molecular characterization using different markers like mitochondrial genes (for example, cytochrome oxidase I gene) or nuclear genes (Internal Transcribed Spacer Region I and II) have provided clearer picture on speciation.

DNA barcoding is emerging as an important tool for identification of organisms. It refers to the technique of sequencing a short fragment of the standardized DNA sequences as markers for the recognition of species. The "DNA Barcode" from a taxonomically unknown specimen and analyzing their performing comparisons with a reference library on barcodes of known species to establish a species level identification (Hebert *et al.* 2003). DNA barcoding has recently been applied to bees that have shown reliable results to be distinguished between different species (Sheffield *et al.* 2009). Previously, integrating DNA barcoding with morphology has also helped in resolving taxonomically difficult groups of bees (Gibbs 2009). Alves (2006) had applied RAPD method molecular characterization on *Tetragonisca angustula angustula* and *T. a. fiebrigi*. He reported that the two bee species were separated with genetic distance values, which were in accordance with that of Nei (1978), who reported it as 0.23 and between the two *T. a. fiebrigi* populations was 0. 0507. Koch (2010) reported that in the taxonomy of the western Malagasy stingless bees (*Liotrigona* spp.), the high degree of morphological similarity between the described species was highly complicated. However, species boundaries of the taxonomically difficult taxa were successfully resolved by application of combined morphology and DNA barcoding analysis. The different species of *Liotrigona* genus were extremely similar in their morphology. Morphologically, *Liotrigona bitika, L. madecassa L. kinzelbachi* sp, and *L. mahafalya* were not clearly differentiated. By application of COI sequence on these species, data as an independent evidence shown as four clusters to be genetically distinct taxa. Ndungu *et al.* (2017) showed clear separation between *Meliponula ferruginea* reddish brown and *M. ferruginea* black into two separate clades by using DNA barcoding. The genetic distance between *M. ferruginea* black collected from

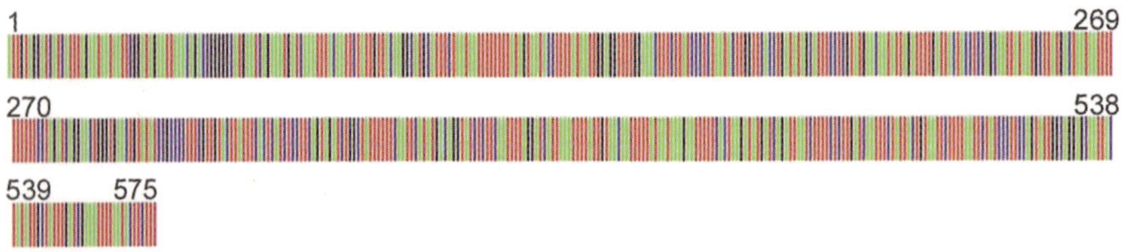

Fig 3.6. DNA barcode of *Lepidotrigona arcifera.*

Fig 3.7. DNA barcode of *Lophotrigona canifrons.*

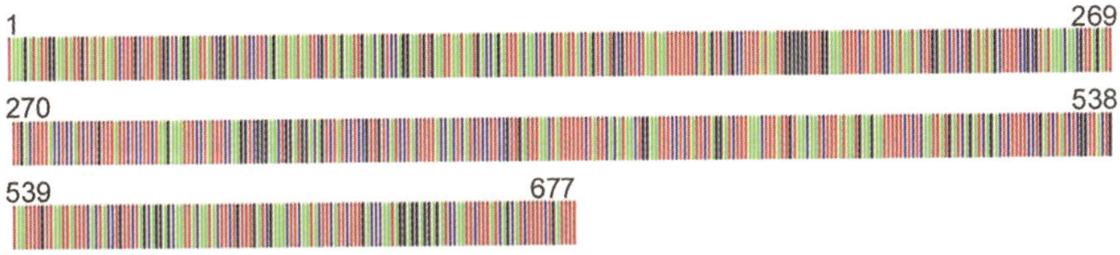

Fig 3.8. DNA barcode of *Tetragonula* sp.

Kakamega and *M. ferruginea* black from Mwingi was low (1.4%), within expected species variation for barcode data based on COI for most animals (Hebert *et al.* 2003) and in accordance with the barcode gap estimate for within species genetic variation (<1.6%). The genetic distance between *M. ferruginea* black and *M. ferruginea* reddish brown collected from Kakamega, in sympatry, was 97.3%.

The identification of stingless bees requires an integration of all morphological, morphometrical and molecular techniques. For some species identification keys are not available while for some, there is no database in the Barcode Library so then it becomes very important to derive the results very carefully while describing a new species or establishing a relationship between separate looking species of stingless bees.

CHAPTER 4

Colony Organization and Life Cycle

Stingless bees like other honeybees are highly developed social insect species that live in colonies. During active season, a normal colony is composed of three kinds of individuals: queen, hundreds to thousand of workers that vary between age as well as species and few drones, which vary in size. In addition, each colony has different developmental stages, viz. eggs, larvae and pupae which are collectively known as brood.

General features of stingless bees

The stingless bees have some peculiar features which makes them different from other corbiculate bees which are described by Wille (1979). It includes

- a vestigial sting which is reduced and is not able to pierce the skin of the intruders.
- The presence of bristles like structures on the internal margins of corbicula, called *penicillium* (in Meliponini).
- A pterostigma present on the fore wing and the marginal cells venation is weak or absent.

Fig 4.1. Stingless bee colony organization (pc prabir).

Table 4.1. Classification of stingless bees

Species	*iridipennis, gressitti, ventralis, arcifera, carbonaria, laeviceps, canifrons*
Genus	*Tetragonula, Lepidotrigona, Lophotrigona*
Sub-family	Apinae
Family	Apidae
Super-family	Apoidea
Sub-order	Apocrita
Order	Hymenoptera
Class	Insecta
Sub-phylum	Uniramia
Phylum	Arthropoda
Kingdom	Animalia

Fig 4.2. General tagmosis in adult stingless bee (Kwapong *et al.* 2010).

- The pollen press is not present as is found in *Apis* genus.
- The hamuli are less (8-16 in meliponini and 4-10 in non-meliponines).
- Maxillary palps absent.
- Wax producing glands on the dorsal part of metasoma.
- The claws are simple in females while bifurcate in males (identifying character).
- Tibial spurs are absent.

Queen

1. The queen is the largest caste in stingless bees identified by a large size, big thorax and distended abdomen not covered with wings. The claws of the hind leg are spread across vertically to pre-tarsus. She is the mother of whole colony.
2. Only one queen is found in a colony but few queen cells can be seen

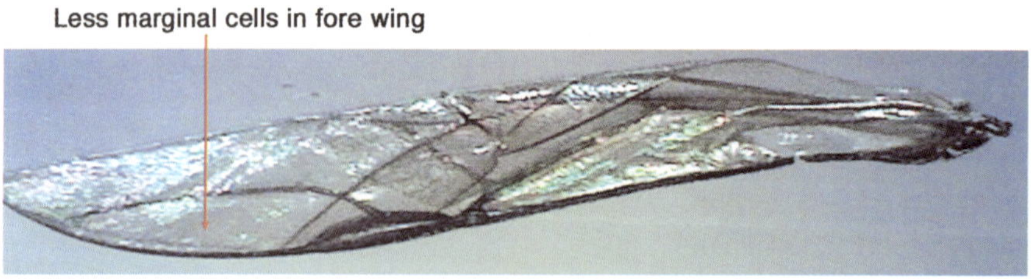

Fig 4.3. Pterostigma on fore wing of stingless bee, *Tetragonula iridipennis*.

during division or supercedure conditions. In some species two or more queens can be observed but all queens are not laying eggs.

3. She leaves the nest only for three times in life. First orientation flight, second nuptial flight and third when issues a swarm.
4. Her function is to lay eggs. The stingless bee queens lays fertilized and unfertilized eggs ranged between 10-100 depending upon food reserves and species.
5. The mating behaviour is variable in different species from single mating to multiple mating. The drones die after mating while the queen bee returns back to the nest.
6. The queens in some species can live up to 3-7 years.

Fig 4.4. Stingless bee queen showing distended abdomen and large thorax.

Fig 4.5. Different castes of stingless bee, *Lepidotrigona ventralis*.

Worker

1. The worker bees make the bulk of the population of the colony.
2. They are imperfectly developed females, morphologically different from the queen in colour and size. The claws of the hind leg of workers are curved like spincers.
3. These perform all the works in the hive except mating and egg laying. When queen is not present, some workers lay unfertilized eggs.
4. The workers feed the colony and the brood is reared by mass provisioning of the cells.
5. Workers have an average life span of 161 days which ranges from 74-286 days depending upon species.

Worker bees do not have stinger but they release a pheromone which induces that attacking and biting instinct in other workers on intrusion or disturbance.

- **Laying workers**: When there is no queen in a colony for a long period, the ovaries of some workers start developing in some species for laying eggs while in other species, the colony slowly collapsed and die.

Drone

1. These are the male members of the colony.
2. In many species males do not perform any duty except mating. In some species drones collect nectar but lacking in corbiculate legs. The claws of the males are bifurcate for better attachment during mating.
3. The proboscis is smaller in males as compared to females. In *M. beechi*, the number of flagellomeres are 10 in queen and workers while in males these are 11 in numbers.
4. In some species the drones leave the colony after few days of their emergence while in some species, the drones live with the colony and mate with the queen.

Males or drones usually stay inside the colony until the age of about 2–3 weeks (*P. droryana*, Cortopassi-Laurino 1979; *M. beecheii* and *M. favosa*, van Veen *et al.* 1997). During their stay within the colony, they have been reported to be involved in building activities, trophallaxis process, nectar dehydration (Imperatriz-Fonseca 1973; Cortopassi-Laurino 1979), incubation, and nest defense (Kerr 1990). According to Nogueira Neto (1997), some activities such as building and nest defense are rather only incidental, even though males are also known to secrete wax. Nectar dehydration could also be considered as a role in their preparation for leaving the nest and therefore may not be counted as a social activity. During their stay in mother colony, the males of *M. beecheii* spent maximum time (>90%) for walking, standing, self-grooming, and stirring (Van Veen *et al.* 1997).

Life cycle of stingless bees

Bees are holometabolous like other insects. The developmental stages of stingless bees are: egg, larva, prepupa, pupa and adult. Queen deposits egg at the base of the cell which contains pollen and nectar, thus facilitating mass provisioning by the workers. After the pupation, the sealed cells can be differentiated on the basis of appearance. The duration of different life stages of different castes in *T. iridipennis* is given below:

Table 4.2. Development period of different castes of stingless bee, *Tetragonula iridipennis*

Caste	Egg period (days)	Larval stage (days)	Pupal stage (days)	Total (days)
Queen	5-6	16-18	25-26	47-51
Worker	5-6	11-13	33-35	49-54
Drone	5-6	19-20	28-29	52-55

The cell construction, provisioning, oviposition and its whole development period from oviposition to adult emergence, i.e. egg period, larval period, pupal period and their growth in terms of length and breadth has revealed an important linkage between different castes. The queen plays an important role in regulating cell construction and discharge of larval food by workers. In some species, egg-laying activity by workers does not take place even in the absence of queen (*L. ventralis*), where the cell construction ceased and the colony diminished day by day (Roopa *et al.* 2015). Wunderink and Sommeijer (1996) observed the same in *Trigona nigrapurpurea*. The brood cells of different stages have different colours. The newly constructed cells are brownish and later become straw coloured with the advancement of age. The workers constructed uniform globular brood cells all connected to each other by pillar like connections in batches of 4 to 6 inside the colony. Single stingless bee worker constructs and complete one cell in 2.0-2.5 hr. The queen inspects the brood cells during construction. After completion, 6-8 attendant workers start filling up of each cell with special honey and pollen food mixture which is regurgitated from their crop. Once that is done and inspected, the queen moved around such filled cells by fluttering her wings in a rhythmic manner. Later she deposits an egg at the centre of larval food, either in vertical or slanting position. Then, within a few seconds, one of the workers started closing the cell with cerumen and finishes it within two or three minutes. The average time required to finish this operation is about two minutes and 16 seconds. Kshirsagar and Chauhan (1977) also reported the same sequences in the development of colony. The entire process of building a brood cell, providing larval food and laying an egg is called provisioning and oviposition process (POP). The eggs are smooth, cylindrical and transparent measuring 1.2 to 1.37 mm in length with a mean of 1.33 mm and width 0.4 to 0.48 mm with a mean of 0.44 mm (Roopa *et al.* 2015). The development stages of generalized stingless bee are given in Fig. 4.6.

The egg hatches in 5.5-6 days with an average of 5.75 days. The larvae is creamy white, apodus and “C” shaped. The length and width of larvae on 3rd, 7th and 12th day of development period shows that it increases with advancement of age of larvae. The larval duration ranged between 11.0 to

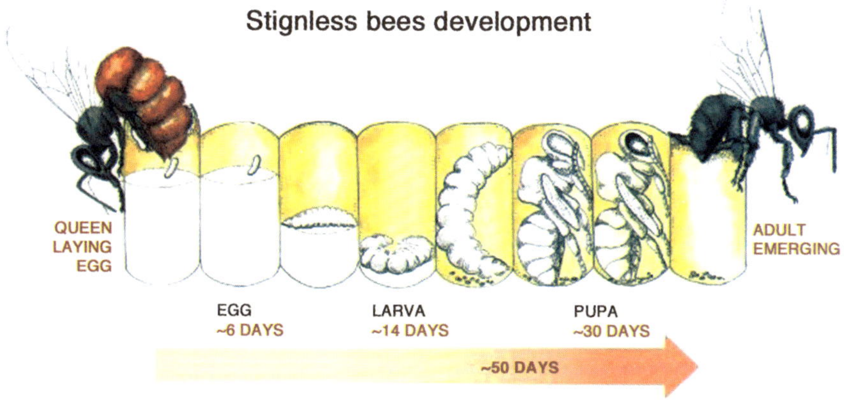

Fig 4.6. Developmental stages of stingless bee (Artwork Gena Cranson).

13.5 days with a mean of 12.7 days. In stingless bees, there is no physical contact between larvae and adult bees since the cell is closed after egg laying. The system of larval feeding in case of stingless bee is called mass provisioning, which resembles the system used by solitary bees. The pupal cells appear transparent as the brood food are exhausted on the removal of cerumen after pupation in the brood cell. The pupae are of exerate type measuring 4.5-4.75 mm in length with a mean of 4.63 mm and width of 2.9-3.1 mm with a mean of 3 mm. The pupal period tooks 33-38 days with an average of 35.3 days. The total period of development from egg to adult required a mean of 53.3 days. Immediately after emergence, the creamish to ash colored young adults started grooming their body and turned darker after eight days. Thereafter, they started removing the waste materials, dead bees, construction of cells, provisioning and guarding at the entrance after 20-25 days and started foraging when they were 35-40 days old. The marked bee under onservation was found dead in about 80-87 days usually inside the nest itself. The dead bees are disposed off by the bees of 20-25 days old.

CHAPTER 5

Social Behaviour of Stingless Bees

Insects are the most dominant organisms on the planet. They are very diverse and form groups from solitary to well established social colonies. Among all the insects, only few orders adopted communal behaviour in their life cycle. The most prominent orders exhibiting true sociality are Isoptera and Hymenoptera. Hymenoptera comprises wasps, honeybees and ants. Only two families of superfamily Apoidea contain highly eusocial species which are Halictidae and Apidae. Most of other bees live solitary life. Halictidae is the most diverse in terms of nesting behaviour as compared to other bees. The nesting behaviour ranges from solitary to highly social in this group.

Social behaviour

Stingless bees are one among the few social insects that have overlapping of many generations in the same nest. Each colony of stingless bee is a well-organized social group having division of labour in terms of nursing, comb building, guarding, food collection and storage maintenance. They have well-developed communication system through different types of courts as well as trophallaxis.

Trophallaxis is a phenomenon of food transmission (exchange of food) in stingless bee and is common between workers and also from workers to queen and drones. It is a sort of communication in regard to the availability of food and water as well as a medium for transfer of pheromone.

The organization of the colony and performance is largely dependent on the quantity and quality of food which the worker bees bring from the flowers. The workers bees are skilled to different aspects of selection for food in terms of the reward and nutritional value through their experiences and innate ability to search the food source. In addition to their innate preferences and previous experiences, foragers can use social information to decide when and where to forage for food. In some stingless bee (Meliponini) species, individual foraging decisions are shown to be influenced by the presence of social information at resource sites (Sommerlandt 2014). Honeybees and stingless bees are the only highly social bees, but more importantly, stingless bees practice a more diverse lifestyles, including obligate necrophagy (Roubik 1982a; Camargo and Roubik 1991), and can recruit workers for different resources such as dead animals, pollen, nectar, mud, resin, water, and searching nests (Roubik 1989). Stingless bees also use a greater diversity of recruitment communication systems (ranging from odour trails to the potential referential encoding of food location) than honeybees and in which, all studied species use the waggle dance (Kerr 1969; Roubik 1989; Dyer 2002).

Social information can be described as any information provided by a sender based on accounts such as behaviour, physical presence or remnant that reduces uncertainty for a receiver. The information can be produced inadvertently or on purpose and may be of different modalities, as for example visual, olfactory or tactile (Danchin *et al.* 2004; Dall *et al.* 2005; Kendal *et al.* 2005; Gruter and Leadbeater 2014). Such publicly accessible information provided by conspecific or heterospecific individuals allows the observer to adaptively change its behavior to gain fitness benefits, such as assessing the location and quality of a food source (Chittka and Leadbeater 2005; Goodale *et al.* 2010, Yokoi and Fujisaki 2011). In bees, olfactory information at the food source can be provided by foragers through anal droplets, gland secretions or footprints originating from glandular epithelia of the claw retractor tendon (Nieh *et al.* 2003; Jarau *et al.* 2004; Barth *et al.* 2008; Wilms and Eltz 2008; Jarau 2009).

Stingless bees use different means to communicate between each other to provide information leading the way to food sources. The communication system of the stingless bees is three dimensional which enable them to forage and locate the food sources in the deep forests and open fields depending on the different species. There are three main methods. The first method is that the scout bee returns and makes a special sound in the nest that makes other bees to fly out and search for the flowers at random. The second method is that the scout bee lays out an odour trail by marking stones and plants on the route with a special scent. Inside the colony, she emits a sound and perform a zigzag dance. When leaving the nest again, she leads a group of recruits to the source by following the trail. The third method is similar to the second, but instead of the odour trail, the scout bee guides a group of workers by means of a pheromone emitted during her flight back to the flowers.

Nieh (2004) described the recruitment in stingless bees very beautifully. He proposed all the three types of communication strategies followed by stingless bees. The efficacy of scent marks released at the point of food source was clealy depicted in *T. corvina* (Boogert *et al.* 2006) and other species of stingless bees (Sanchez *et al.* 2011). They emphasized the importance of olfaction as the major sensory modality. It is reported that the cause may either be due to a bias in sensitivity for visual and olfactory stimuli which is a difference in the stimuli's action range or a true (innate) preference of the bees for scent marks over visual stimuli. Odour marks can act effectively over larger distances than visual cues with the exception of non-volatile long chained epi-cuticular hydrocarbons. In the stingless bee, *Tetragonisca angustula* (Latreille) the estimates of spatial visual resolution that have 5-6 mm body length is as about the size as *T. corvina*, suggests that objects within the size of conspecifics should be visually perceptible at relatively short distances of approximately 10 to 20 cm (Zeil and Wittmann 1993). In contrast, odour marks deposited by foragers can be sensed by conspecific stingless bees even over longer distances up to 10 to 20 m (*Melipona panamica* Cockerell: Nieh 1998; *Scaptotrigona aff. depilis* (Moure): Schmidt *et al.* 2003).

Mechanisms of information transfer

The different mechanisms employed in stingless bee communications are through contact and visual communication, olfactory communication and acoustic communication.

Trophallactic contact which literally means food exchange may be the most primitive form of meliponine communication (Hart and Ratnieks 2002) and could provide information about food quality and odour (de Marco and Farina 2003). For example, Hrncir *et al.* (2000) have shown that the number of trophallactic (food exchange) contacts between *M. quadrifasciata* and *M. scutellaris* recruiters and potential recruits increases before the recruit successfully reached the feeder.

Outside the nest visual communication plays a role in recruitment orientation. Slaa *et al.* (2003) have shown that foragers of stingless bee species, viz. *Trigona amalthea*, *Tetragonisca angustula, Partamona cupiraii*, and *Oxytrigona mellicolor* tend to display local enhancement and orientation towards the food source where visual presence of nest mates are readily available. It is known that the foragers of several species may also methodologically guide experienced foragers, at least for some of the distance to the food source and visual tracking of the leader is speculated as the primary orientation mechanism (Esch 1967; Kerr 1969). Esch *et al.* (1965) reported to have observed *M. quadrifasciata* foragers perform a zigzag flight pointing towards the direction of the feeder but the nest mates usually did not cross a distance of 30–50 m (Esch 1967). Kerr (1969; 1997) theorised that the whitish reflective abdominal hairs of some species may be able to facilitate orientation in the direction of experienced foragers as they move out from their respective nests.

Several odour sources like food odour, olfactory signals, cues deposited by foragers, locale odours and odours of the environment surrounding the food source can influence forager orientation in socially active bees (von Frisch 1967; Goulson *et al.* 2000; Goulson *et al.* 2001; Dyer 2002). Among all bees, stingless bees are the only bees known to produce long odour trails beginning from the nest location and extending to the food source (reaching a reported 900 m in *T. amalthea,* Kerr 1960). Although odour trails usually lead to nectar or pollen resources, obligate necrophages such as *T. necrophaga* reportedly use odour trails to indicate their food sources (Roubik 1982a).

There is a wide scale of strategies, ranging from complete odour trails which may begin from the nest vicinity and extend to the food source or even short odour trails that may extend only for a short distance away from the food source but directed towards the nest (Nieh *et al.* 2003a) and also, distinctive odour-marking of the food source alone (Nieh 1998b; Hrncir *et al.* 2004). In many meliponine species odour trail marking has been reported positively (Lindauer and Kerr 1958, 1960; Kerr and Cruz 1961; Kerr *et al.* 1963; Kerr and Esch 1965; Kerr 1960, 1969, 1972, 1973, 1994; Blum *et al.* 1970; Kerr *et al.* 1981; Kerr and Rocha 1988).

Complete odour trails

Stingless bees are known to produce an odour trail that consists of odour droplets deposited on vegetation periodically between a food source and the nest. The distance between odour marks ranges from 1 m to 8 m in different species (Kerr 1969).

Short odour trails

Stingless bees such as *M. rufiventris* and *M. compressipes* deposit a short odour trail that may consists of one to three putative "urine" marks (anal droplets) and extending up to 8 m from the food source (Kerr 1994). Nieh *et al.* (2003a) reported that *T. hyalinata* foragers produced short odour trails that extended up to 27 m from the food source towards a nest 146 m away.

Aerial odour trails

Kerr (1969, 1994) hypothesized that foragers of some species like *Partamona helleri* deposit aerial odor trails and create "odour tunnel" as they continue their flights to the food source. Kerr (1994) proposed that this mechanism could function during windless conditions.

Odour mark sources

The odour marks of meliponine, have several glandular sources (Cruz-Landim *et al.* 1998) and may also consist of cues such as anal droplets. The anal droplets are clear droplets of fluid excreted from the anus of foragers, usually after they have finished their feeding. The phenomenon have been described in several species: *M. rufiventris* and *M. compressipes* (Kerr and Rocha 1988), *M. favosa* (Aguilar and Sommeijer 1996, 2001), *M. panamica* (Nieh 1998b), and *M. mandacaia* (Nieh *et al.* 2003d) and are generally produced in greater number at more dilute food sources for which foragers do not recruit (*M. mandacaia*) or which lead foragers to greatly reduce their rate of recruitment (*M. panamica*).

Odour-mark deposition

The stingless bees can deposit odours through several significant methods. The mandibular gland secretions may be deposited by rubbing the substrate with the mandibles (Kerr 1969), and tarsal gland secretions may be secreted by walking on the substrate (Jarau *et al.* 2002). While depositing odour marks on surrounding foliage, foragers generally land briefly (average of 1.4 ± 0.7 s for *T. hyalinata*, Nieh *et al.* 2003a) and rub their mandibles (Lindauer and Kerr 1958; Kerr 1972; Kerr and Rocha 1988) and tongue (*T. hyalinata*, Nieh *et al.* 2003a) against the substrate. Foragers can also groom while depositing odour marks, but it is unclear if grooming deposit odours (Kerr 1994; Nieh *et al.* 2003d).

Acoustic recruitment communication

Sound production is an intriguing aspect of meliponine recruitment behaviour. Bees do not possess tympana and are unable to hear the pressure component of sound (Michelsen *et al.* 1986b). However, the Meliponini bees have the ability to detect the particle-velocity component of sound (Kirchner 1994; Michelsen 2003) and sound vibrations transmitted through the substrate (Nieh and Tautz 2000). In recruitment of stingless bees, sound plays important role in activating nest mates or mates from the same nest that have previously experienced the feeder (reactivated foragers) to go out and forage (Lindauer and Kerr 1958; Esch 1967). In addition, correlations have been found between recruitment sounds and the quality and location of the food source (Esch *et al.* 1965; Nieh and Roubik 1998; Aguilar and Briceno 2002; Hrncir *et al.* 2002; Nieh *et al.* 2003c). In *Meliponula (Axestotrigona) ferruginea tescorum* and *Nannotrigona testaceicornis*, Kerr (1969) reported a certain form of acoustic chorus produced by the bees. In these species, when a worker, having discovered a source of food, enters the hive, it produces a characteristic sound. Immediately thereafter the bees closest to it begin to produce the same sounds, and then others join in, and within less than 1 min the entire hive will be buzzing, and all available bees leave the hive looking for food. Kerr *et al.* (1963) also reported that *M. quadrifasciata* foragers produce sounds on food sources. In *M. panamica,* foragers produce sounds upon departing from a feeder, and the rate of sound production is positively correlated with sucrose concentration, suggesting that feeder sounds may be related to forager perceptions of food quality (Nieh 1998b). Esch *et al.* (1965) discovered that some stingless bees (*M. seminigra merrillae* and *M. quadrifasciata*) may encode the distance to the food source in the duration of recruitment sound pulses. The foragers of both species increased the duration of sound pulses as and when distance to a food source increased.

Hrncir *et al.* (2002) have shown that recruitment sounds in stingless bee can vary with forager motivation. *Melipona seminigra* foragers reduced the amount of recruitment sound produced whenever food quality decreased by decreasing the pulse duration and increasing the interpulse interval (the time between two sound pulses). Limited sucrose solution flow also led to shorter pulses and longer interpulses than unlimited flow, and pulse duration increased with increasing foraging duration (Hrncir *et al.* 2002). Nieh *et al.* (2003c) reported that food quality (sucrose concentration) has a similar effect on *M. mandacaia* and *M. bicolor* recruitment sounds, decreasing the pulse duration and increasing the interpulse duration as sucrose concentration decreases.

Thus, the foragers of stingless bee may employ multiple sensory modalities (touch, vision, olfaction, and audition) to transfer information in regard to the existence of a resource and, in some cases, to communicate specific 3-dimensional location among themselves. These modes of information transfer mechanisms allow the colony to allocate the foraging resources collectively and judiciously (Roubik 1989; Biesmeijer *et al.* 1998, 1999a; Biesmeijer and Ermers 1999; Biesmeijer and Slaa 2004). Like honeybees, stingless bees act as a super organism whereby they practice with decentralized control of foraging that is established upon the decision of each individual foragers (Seeley 1983, 1985, 1989a, b; Biesmeijer and de Vries 2001; Biesmeijer and Slaa 2004). Therefore, the transfer of recruitment information should be reliable (Nieh 1999), multimodal communication may furnish information inside the often noisy and crowded nest (Holldobler and Wilson 1978, 1990) that enable foragers to alert nest mates reliably about the existence of a good resource nearby.

■

CHAPTER 6

Stingless Beekeeping Equipment

Before starting beekeeping with stingless bees, one should have the familiarity with different equipments. The different equipments for successful stingless beekeeping are bee hive, stand, hive tool, knife and swarm basket. These are the common equipments which can be availed easily from the practitioners or beekeepers.

Stingless bee hive

Stingless bee hive is the most important as it ensures the colony establishment for years without loss of brood and bees during extraction. Different researchers tried various types of bee hives for the domiciliation of stingless bees. Traditionally, from the last many centuries people are using wood log hives, bamboo hives, wooden plates, bottle gourd hives etc. With the advancement and interest of researchers in stingless bee's nest architecture, development of hives started. The hives were developed for different species (Heard 1988; Sommijier 1999; Lindemann and Arya 2002; Punith 2014) but in India no scientific hive is still developed for the rearing of *Tetragonula* spp. and *Lepidotrigona* spp. (Chauhan and Singh 2019).

It is possible to divide the presently existing and the most frequently used hive models into two groups–horizontal and vertical hives. Traditionally, horizontal hives are the most commonly used hives in Brazil and Mesoamerica (Nogueira-Neto 1997; Sommeijer 1999; Cortopassi-Laurino *et al.* 2006). Horizontal hives are somewhat similar to hollow logs, placed horizontally (Vilamueva-G *et al.* 2005), and are the common method of keeping stingless bee nests (Nogueira-Neto 1997). Horizontal hives may be completely hollow, without any internal division, or more elaborately, with internal divisions to separate the nest from the honey storage space (Nogueira-Neto 1997; Sommeijer 1999). The vertical hive, Baiano or Nordestino, Uberlândia, PNN, the UTOB and the INPA hive models are widely used in Brazil and in other countries as well, with or without local adaptations (Quezada-Euán and Gonzalez-Acereto 1994; Kerr *et al.* 1996; Nogueira-Neto 1997; Sommeijer 1999). The many hive models in the meliponiculture are directly indications on the vast diversity of stingless bee species present in these regions (Venturieri *et al.* 2012), and also based on personal and/or cultural preferences, many beekeepers also prefer to create new hive models by themselves even when the more known models (i.e. INPA, Embrapa, PNN, UTOB) are available. The number of beekeepers using vertical hives is growing despite the horizontal models still being more widely used (Venturieri 2008a). The vertical hive follows the natural brood comb pattern in tree trunks, divided into two main modules, the base chamber (to shelter the nest) and the upper chamber (to store honey and pollen; Portugal-Araújo 1955). The creation of a specific honey storage compartment enables a faster and cleaner harvest, with minimum nest damage, which is a great advantage of this hive model (Venturieri *et al.* 2003; Venturieri 2008a). However, actual comparisons on the advantages of each type of hive with regard to biological parameters, such as hive occupation and thermoregulation, for instance, are scarce but in general, the type of hive used for rearing different species and the resultant breed from such hives show clear differences in the health and development of the colonies (Quezada-Euán and Gonzalez-Acereto 1994).

A hive box with solid wooden structure was designed by Heard (1988). This is sawn into halves for rearing *Trigona carbonaria*. The box has 28 cm × 21 cm base and 21 cm height and the entrance hole of 1.2 cm was drilled into bottom half and the boxes were painted. To provide good platform, the bottom board may be extended. He used timber Oregon pine (Douglas fir) with thickness of 25 or 30mm, which was durable and provides good insulation. Good quality glue and galvanised nails must be used for manufacturing of boxes and he used bees wax for joining the boxes. Similarly, Sommeijer (1999) tried working on rearing of *Melipona fevosa* and developed 'Utrecht University-Tobago Hive' for stingless bee rearing in Tobago. The hive consists of two main parts, namely the brood chamber and the honey chamber which rested on a wooden bottom tray that is surrounded by a rim. The chambers fit tightly into the rim of the bottom tray. When both are placed on the tray, the wooden wall that separates the brood chamber from the honey chamber, has an aperture where the bees can pass unhindered between brood and honey chamber. The hive entrance is on the other side that is away from the brood chamber and as such, minimum disturbance is maintained in the brood nest and significantly reduces the risk of an invasion by phorid flies. The hive should be constructed in such a way that it facilitates quick, easy removal and replacement of the honey chamber. For inspection purpose, the brood chamber should be a separate unit and should be just big enough to satisfy a good-sized brood nest. The size of the particular hives that were constructed were adapted to scientifically rear *Melipona favosa*, a relatively small stingless bee with small colonies.

Lindeman and Araya (2012) showed stingless bees use longer entrance to protect the hive from intruders. Based on this, they constructed another type of scientific stingless bee box. The bee box consists of brood chamber, stackable honey chamber and a lid. The different parts fit together using metal dowels. In order to create appropriate size and shape, pieces of wood were placed in each corner of brood chamber. In this box, each honey chamber was provided with thin sticks that slide across the top and thereby provide a place for bees to attach the honey pots. The brood chamber too was filled with pieces of wood in each corner for giving appropriate size and shape. The inner internal diameter of entrance hole was less as compared to the outer external opening. A vertical inner channel was drilled for protection and creating longer channels.

Kumar *et al.* (2012) studied the stingless bee keeping in the tribal areas of Western Ghats in Tamil Nadu, where they reared in bamboo pole which would help in improving the new modern techniques in order to get pure, clean and uncontaminated honey.

Punith (2014) studied on hive designs and multiplication methods of stingless bee *Tetragonula iridipennis*. He took wooden hives with three dimensions, viz. small (15 × 10 × 10 cm^3), medium (30 × 10 × 10 cm^3) and large (30 × 15 × 10 cm^3). Among three hives, medium sized hive was well suited for rearing and recorded maximum brood, food store and weight gained in them. Each hive was made of wooden plank of 2.5 inch thickness which made box steady and long lasting.

Pallavi (2016) studied seasonal performance of stingless bees as influenced by different hive designs. The hives used were of four different types: wooden hives measuring 30 × 10 × 10 cm^3, bamboo stem hive measuring 30 cm in length and 10 cm in diameter which were vertically split and joined, plastic hive made of insulated plastic pipe of 30 × 10 cm^2 which was insulated using another pipe of wider diameter and the insulated material was saw dust and the last hive used was Kerala hive which was a rectangular wooden hive measuring about 20 × 12 cm^2. As per their studies, wooden hives measuring 30 × 10 × 10 cm^3 was found to be best suited for rearing of stingless bees. Leao *et al.* (2017) developed and analysed the horizontal hives for hive occupancy, brood and honey development and breeding of *Scapatrigona postica.* They observed the division process was easily carried out in the vertical hive, once this hive model has subdivisions and split rods. Brood comb separation occurred naturally in the vertical hive, without major interventions, because the use of sticks between the nest and the upper nest facilitated separation and minimized damage. Another advantage of the vertical hives over horizontal hive is the facilitation of honey extraction. In vertical hives, the honey is stored in a separated compartment (the super), which can be easily separated from the hive without disturbing the colony (Venturieri 2008a). Singh (2016) has reported that the beekeepers of Naga tribes from Nagaland, India used traditional hives, which are not scientifically standardized. He mentioned in his report that bee-keepers use Log hive, bee box and rectangular wooden box hives. Reportedly, log hives having 43-123 cm long and 11-38 cm in diameter were used at different districts of the Nagaland for *T. iridipennis* and *T. laeviceps.* Central Bee Research and Training Institute, Pune recommended wooden boxes of 10 × 12.50 × 10 cm size with a glass top covered by a wooden lid for rearing *T. iridipennis* (Percy 1989). Coconut shells, small earthen pots and bamboo bits were being used in Kerala for rearing stingless bees from olden times and people are using the same even today. Wooden hives of size 3 × 14 × 15 cm were found superior to bamboo hive and earthen pots for rearing *T. iridipennis* (Raakhee 2000).

Chauhan *et al.* (2019) tried different types of hives for scientific stingless bee rearing (Fig 6.1). During field survey of different parts of the country,

Single chambered hive Three chambered hive Bamboo hive

Nine chambered hive Five chambered hive

Fig 6.1. Different types of stingless scientific bee hives.

they realized farmers were using the traditional hives (log hive, bamboo hive etc) around the whole country which are having constraints during honey extraction and pollination deployment. The different hives developed by them were three storey vertical hives, single chambered hives, five chambered hives, nine chambered hives and modified bamboo hives (not published) for stingless bee rearing. The merit of these hives is easy extraction of honey with minimum disturbance to brood areas in the colony and easy portability to the field areas for pollination as compared to traditional log hives.

Parts of stingless bee hive

a. **Stand:** It supports the bottom of the hive and keep the bee colony safe from enemies.
b. **Bottom board:** It is the floor of the hive having an entrance for the bees.
c. **Central chamber:** It is a rectangular box/square in which the bees rear their brood and surround the brood with pollen and honey in *T. iridipennis* and on back side of brood in *L. ventralis*.
d. **Inner cover:** A board which acts as a partition between top cover and brood chamber.
e. **Top cover:** It rests loosely over the hive which acts as a roof placed over inner cover.

Equipments used in meliponiculture (Fig. 6.2)

Hive tool: An iron strip or chisel used for opening of hive and its cleaning.

Stingless bee hive on stand

Hive tool

Bee brush

Bee veil

Hive stand with ant barrier

Fig 6.2. Different equipments used in meliponiculture

Bee veil: It is used to prevent bee bites on face and neck.

Uncapping knife: Large size knife used to remove the propolis from the sides of the hive during extraction of honey.

Bee brush: To brush the bees from brood and honey during division of colony and extraction of honey.

Gloves: These are used to protect hands from bee bites.

Ant barriers: These are the plastic or clay bowls filled with water used to prevent the robbing of nectar and brood of ants. ■

CHAPTER 7

Meliponiculture: Handling of Bee Colonies

Meliponiculture is the branch of science which deals with the rearing as well as raising of stingless bee colonies, their management for building economy and conservation of biodiversity. The domestication of stingless bees is an ancient practice in the Americas, carried out mainly by traditional communities (Cortopassi-Laurino *et al.* 2006). Generally, these communities still keep the stingless bees in a rudimentary way (Villanueva-G *et al.* 2005; Reyez-González *et al.* 2014), although recently, attempts have been made to enhance general bee keeping techniques and practices (Contrera *et al.* 2011; Venturieri *et al.* 2012). However, stingless beekeeping remains an essentially informal activity across developing countries where technical knowledge is scarce, and lack of standard management practices (Jaffe *et al.* 2015). For meliponiculture, some few points may help in better management of bees

Fig 7.1. A scientific Meliponiary at School of Agricultural Sciences and Rural Development, Medziphema, Nagaland.

for honey and pollination under protected conditions. The key to successful beekeeping lies mainly in three things:

A. Good meliponiary site
B. Appropriate bee species
C. Proper management

Selection of good meliponiary site

- The selected site where stingless bee colonies are to be reared should have sufficient bee flora which can provide forage for a major part of the year.
- The Meliponiary should be easily accessible by road.
- The availability of fresh running water is necessary.
- The Meliponiary should have natural or artificial wind breaks to protect bees from cold winds.
- The site should receive morning and afternoon sunshine and during summers, provision of shade should be provided to the hives.
- Stingless bee shed can be prepared to protect the bee colonies from rain.

Selection of good bee

Stingless bee keeping can be taken up with meliponiary or available species in the area. In higher altitude regions, *Lepidotrigona* spp. should be preferred since they adapt to chilly climate better while in subtropical regions, *Tetragonula* spp. can be considered for rearing in homestead Apiary.

Proper management

The success of stingless beekeeping also depends upon proper understanding of the particular bee behaviour and manipulating the colonies accordingly. For appropriate manipulation of bee colonies, periodic examination is required with dedication.

Handling of bee colonies

For management of stingless bees, examination of bee colonies form one of the most important aspects.

Why to examine stingless bee colonies: A bee colony is examined to check its activities condition and to determine its requirements at a particular time since these vary during different periods of the annual cycle of a bee colony. When a bee colony is opened, one should always make the following observations:

a. Whether a bee colony has sufficient food or it needs artificial feeding (*Tetragonula* spp. only).
b. Whether the queen is present or not. Do not disturb the hive too much as the queen may sometimes, stay in central brood portion for hatching and cannot be observed. In *Lepidotrigona* spp. the whole colony is inside an involucrum and so queen cannot be seen. But the presence of queen can be judged from the developing size of the involucrum.
c. Whether there are any of the enemies or diseases in the colony.

Stingless bees like most bees, do not like interference as it affects their normal working system/routine. Therefore, the colonies should be disturbed as little as possible or only as and when required but to the minimum. The observed minimum time for development after division of stingless bee colony is 30–50 days since the new adults emerged from the cells only after a mean period of 40-50 days.

Requirements for examination

The required tools for proper management include hive tool, bee veil, gloves, meliponiary record register, measuring scale or grid, plastic vials for collection of pests/ disease if any, etc.

Maintenance of meliponiary records

Observations: It is very important to keep the record of every colony and enter your observations whenever the colony is examined. The Proforma for recording observation is as under:

Table 7.1. Data record sheet of stingless bee, *Tetragonula iridipennis*
Date of observation: ....*Tetragonula iridipennis*

Colony number	Brood area (L × B)	Pollen area (L × B)	Honey (g)	Queen present or not	Remarks

Table 7.2. Data record sheet of stingless bee, *Lepidotrigona ventralis*
Date of observation: .... *Lepidotrigona* spp.

Colony number	Involucrum area (L × B)	Pollen area (L × B)	Honey (g)	Queen present or not	Remarks

Procedure

1. Use bee veil, gloves and hive tool while opening a bee colony for inspection.
2. Wear bee veil and do not stand in front of hive entrance.
3. With the help of hive tool, lift gently and remove the top and inner covers.
4. Look for the queen and be careful while carrying out the inspection activities in order to avoid any damage to the brood or queen.
5. Check whether the brood is healthy or not.
6. Always be careful to avoid damage to hive entrance especially in *Lepidotrigona* spp.

CHAPTER 8

Flora of Stingless Bees

Stingless bees being the large group of Apidae are still the least known among all social bees, probably because of their tropical distribution and less focus by researchers. Since a long time, the traditional people have conserved these important bees and now, the researchers are also showing immense interest to explore these bees. Stingless bees are dispersed throughout most parts of the world and tropical and subtropical regions of India. Stingless bees form an important group of pollinators in agricultural and natural ecosystems. Plants are an essential component of stingless bee's life as their brood rearing completely depend on the availability of floral resources. Likewise, an abundant supply of nectar and pollen producing plants within the flight range of the bee colony is vital for production of honey. Bees and flowering plants have a long relationship, developing together over the time in a way that benefited both the plants and the bees and enabling both to survive better than they would have done alone. Worldwide, about 40,000 plant species are of value to bees while about 4,000 are the source of most of the worlds' honey. Among the important economic crops, bees have been shown to increase the yield of oil seeds, stone and pome fruits, berries, vegetable and almost all fruit and seed crops (Partap 1997).

A total of 39 pollen taxa were identified in dry chaco forest of Argentina. The collection of pollen by stingless bees was focused on a few pollen taxa namely *Prosopis, Castela coccinea, Maytenus* and *Capparis*. These taxa, together with *Ziziphus mistol* and *Pisonia zapallo* were also identified as vital nectar sources while pollen collection varied according to season with summer being the most diverse when *G. argentina* incorporates herbaceous plants in their diet. The spectrum of pollen collection for *G. argentina* is akin to that of other *Trigonina* bees wherein the main plant species visited include some large shrubs or trees which have small and clustered flowers. The stingless bees indicate an organised foraging behaviour since food pots mainly contain huge amounts of monofloral pollen loads from only a few species (Vossler *et al.* 2010). In San Daniele, 172 botanical species belonging to 55 different families are important for bees. Antonini *et al.* (2006) recorded 22 flowering plant species visited by *Melipona quadrifasciata* for pollen and nectar collection. The plant species belonged mainly to Myrtaceae, Asteraceae and Convolvulaceae. *Plebia schrottkyi* foraged on 52 plant species belonging to 26 families. The most abundant families were Euphorbiaceae, Leguminosae, Compositae, Anacardiaceae, Liliaceae and Myrtaceae (Ramhlo *et al.* 1989; Cortopassi and Ramhlo, 1988). A total of 73 plant species were recorded as bee flora belonging to 19 families mainly represented by Compositae and Cruciferae, in addition to Citrus (lemon, orange, clementine) and fruit trees (cherry, almond, prickly pear, apple, pear, apricot, peach). Total 107 plants were recorded as honeybee flora in the semiarid region of Paraiba, Brazil (Silva *et al.* 2012). Bees visited 66 species out of the total 97 bee plant community evaluated in Salvador, Bahia, Brazil. *Waltheria cinerescens* and *Byrsonima microphylla* were the most abundant species (Viana and Kleinert 2005). At quartzitic Ouro Branco Mountains and Ouro Preto fields in Brazil the bees visited 46 flowering plant species, the most visited plants were Asteraceae and least belonged Solanaceae (Araujo *et al.* 2006).

In Dharwad, Karnataka, about 140 plant species were recorded as bee forage sources where 14 are identified as major food sources. The recorded bee flora of Punjab, India comprises 172 plant species belonging to 53 families where a total 35 species were found to be minor source of pollen and the rest 36 species act as major pollen source in Coorg (Suryanarayana 1966). Chauhan and Singh (2018) recorded 89 plants as stingless bee flora which sustains the stingless bee population during different seasons all round the year in Nagaland.

Melissopalynological studies

There is a distinction between pollen analysis and palynological analysis. Pollen analysis is where the presence of pollen grains in honey, pollen loads, bee bread and propolis/geopropolis are analysed scientifically. On the other hand, considering additional structured elements notified in these products, as bacteria, spores and hyphae of fungi, yeast, oil, wood, hairs, amounts of organic material, etc., it is a palynological analysis (Barth 2013). Visiting flowers for nectar obtaining, Meliponinae were sometimes considered to be specialists, while producing monofloral honeys, and on the other hand characterised as generalists, producing heterofloral honeys. It depends upon several factors such as the blooming plant species and the disposable number of flowers, the content of sugars and water in nectar, the weather as well as the bee species. Studies on honey samples of Meliponinae in the state of Sao Paulo (Iwama and Melhem 1979; Ramalho *et al.* 1985; Guibu *et al.* 1988; Ramalho 1990) reported that bees visit a number of flowering species but selectively those belongs to Alchornea, Arecaceae, Eucalyptus, Moraceae, Myrtaceae, Petroselinum and Schinus. A total of 23 genera and 15 families were identified from melissopalynological analysis of 33 pollen types collected from 11 *Melipona capixaba* hives, namely Fabaceae (7), Myrtaceae (3), Solanaceae (3), Arecaceae (2), Asteraceae (2), Euphorbiaceae (2), Melastomataceae /Combretaceae (2), Rubiaceae (2) and Sapindaceae (2) showed the highest pollen richness. The most frequent pollen types (>45%) were Eucalyptus, which generated great similarities between the samples (Luz *et al.* 2011). Bhargava *et al.* (2009) during the research experiments on presence of pollen grains in honey samples recorded different sources of pollen in honey in Karnataka.

Pollen from nine plant taxa *Ageratum houstonianum* (Compositae),

Bursera spp. (Burseraceae), *Coffea arabica* (Rubiaceae), *Muntingia calabura* (Tiliaceae), *Heliocarpus donnell* (Tiliaceae), *Miconia argentea* (Melastomataceae), *Pimenta dioica* (Myrtaceae), *Quercus* sp. (Fagaceae) and *Vernonia* sp. (Compositae) had greater values of importance to *Scaptotrigona mexicana* and *Apis mellifera* in northern Puebla State (Ramirez and Martinez 2007). In a semiarid region at the state of Bahia, pollen analysis of 11 honey samples were obtained from *Melipona mandacaia* and studied accordingly (Alves *et al.* 2006). *Piptadenia rigida* (Fabaceae-Mimosoideae) was the dominant nectariferous pollen in six samples, and *Ricinus communis*, a nectariferous plant species, in one sample. *Piptadenia moniliformis* pollen grains were dominant inside one sample (97.6%) from Paraguassu, Bahia (Santos – Junior and Santos 2003), and a Euphorbiaceae species (51%) in *Trigona spinipes* honey at São Cristóvão, Sergipe State. In the south and southwest of the Buenos Aires Province, Argentina, a melissopalynological map was sketched using pollen analysis data pertaining to 127 honey samples from the Pampa, Espinal (the Calden District), and Monte de Llanuras Mesetas ecoregions. The samples were collected over the period 1992-2002 and was prepared by Valle *et al.* (2007) which revealed that *Eucalyptus* spp. (*E. camaldulensis, E. globulus, E. viminalis*), *Melilotus* spp. (*M. albus, M. indicus* and *M. officinalis*) and *Helianthus annuus* were dominating pollen types. Asteraceae and Fabaceae were the most represented families.

Eighteen pollen types, belonging to 16 plant families, were found on *T. angustula*. Among them, the more frequent pollen types were Meliaceae, Myrtaceae (*Eucalyptus* spp.), *Piper mollicomum* (Piperaceae), *Schizolobium parahyba* (Caesalpiniaceae) and *Tibouchina granulosa* (Melastomataceae) (Morgado *et al.* 2011). Sanchez Chaves *et al.* (2002) studied the diversity of pollen sources in *T. collina*, *S. pectoralis* and *Tetragonisca angustula* and reported *Bursera simaruba*, *Sapindus sapanaria* and a type of Moraceae as important pollen types. Studies on pollen loads in Pozo Azul and Costa Rica disclosed that, *T. angustula* colonies collected pollen from 46 and 28 plant species respectively and out of which 18 were common to both the places. *M. beecheii* colonies collected pollen from 19 plant species (Biesmeijer and Slaa 2004).

Viraktamath *et al.* (1999) accomplished comparative studies on pollen sources of *Trigona* sp., *A. mellifera* and *A. cerana* at Dharwad. They found that *Trigona* sp. exploited 17 species of plant and the pollen sources were strictly different from *Apis* species. Among all different plant species, *P. pterocarpum* was the only common and important source for all the species. The trophic resources of pollen load of Meliponinae bees were analyzed by Carvalho *et al.* (2001) where he identified pollen of *Eucalyptus* and *Psidium* as the dominant pollen. In yet another study, Pick and Blochtein (2002) analysed pollen loads of *Plebeia saiqui* and percentage of the pollen types in the samples was estimated to be 20% Asteraceae, 17% Myrtaceae, 15% Meliaceae and 10% Euphorbiaceae. Dejtisakdi *et al.* (2005) also analysed the acetolized pollen from the pollen loads collected by *T. collina*. It was proven that 21 flowering species served as food plants of the bees based on the pollen morphology.

The pollen load numbers of 3, 16, 18 and 20 were recorded for *Garcinia speciosa*, *Mimosa pigra*, *Engelhardtia spicata* and *Helixanthera* sp., respectively. A melissopalynological study made on 21 samples of commercial Uruguayan honey found 66 pollen types. All the different honeys typically contained pollen from the native *Baccharis* spp., *Eryngium* spp., *Scutia buxifolia* and *Salix humboldtiana* together with pollen from the introduced *Eucalyptus* spp., *Lotus corniculatus*, *Trifolium pratense*, *T. repens* and *Echium plantagineum* (Daners and Telleria 1998).

In Venezuela, pollen spectra were obtained for honey samples taken from the nests of 48 *Melipona* spp. and 20 other stingless bee species of 23 different locations. The order of abundance for dominant pollens in the honey of *Melipona* spp. was *Mimosa pudica*, *Scrophulariaceae*, *Machaerium, Avicennia, Myrtaceae, Mimosa scabrella, Cassia, Myrcia, Piper, Philoxerus, Xanthoxylum, Alternanthera* and *Astragalus* (Vit 1994) *Allophylus, Baccharis, Citrus, Cecropia, Campomanesia, Eucalyptus, Matayba, M. scabrella, Paspalum* and *Vernonia* represented the pollen spectrum in honey samples from the state of Paraná with the major occurrence of *Eucalyptus* (Ramalho *et al.* 1991).

The pollen present in different honey of Meliponinae represented a very rich and heterogeneous spectrum of plant taxa, especially Alchornea, Eucalyptus, Petroselinum and Schinus (Iwama and Melhem 1979; Ramalho *et al.* 1990). The analysis of mangrove honey (Barth and Luz 1998; Luz and Barth 2001) revealed a pollen spectrum of *Eupatorium*, *Gochnatia*, *Mimosa bimucronata/M. caesalpiniaefolia*, *M. pudica* and Sapindaceae. The dominant pollen grains in honey sample of the "uruçu" Meliponidae from Bahia, was Eucalyptus and Psidium, together with Bauhinia, Caesalpiniaceae and *Mimosa verrucosa* (Carvalho *et al.* 2001).

In India, eight pollen types from Uttarakhand and seven from Uttar Pradesh were the predominant pollen types in honey. The investigation further revealed that in addition to already known bee forage (e.g. Brassica, Coriander and Litchi), some other plant species including Eucalyptus, Callistemon, Myrica, Rumex, Erigeron are also heavily availed as pollen and nectar sources by honeybees (Dutta *et al.* 2008).

Ramanujam *et al.* (1993); Ramanajun and Kalpana (1992) also analysed the pollen content from honey samples collected from *T. iridipennis* in Hyderabad, India. The honey was found to be unifloral origin with *Prosopis julifera* as the predominant pollen type and *Rotala densiflora*, *Peltophorium pterocarpum* and *Cocos nucifera* as important secondary pollen types. The other constituent minor pollen types included *Eucalyptus globosa*, *Cyanotis* sp., *Leucaenea leucocephala*, *Tamarindus indica*, *Delonix regia*, *Azadirachta indica*, *Lawsonia alba*, *Ageratum conyzoides* and *Loranthus longiflorus*. One analysis of honey and pollen samples from *T. angustula* have shown a total of 105 pollen types. The plant species were classified as yielding of nectar and pollen both, nectar only and pollen only (Sousa *et al.* 2002).

Melissopalynological studies on 21 honey samples of *T. iridipennis* collected from 10 localities of Himachal Pradesh by Sharma and Gupta (1997) divulged the presence of dominant plants as *Brassica* sp., *Adathoda* sp.,

Clematis sp., *Mussendra* sp. and *Helianthus* sp. An analytical study on pollen in North Eastern region also indicated both unifloral and multifloral types of honey samples (Singh 1999). Botanical studies in melissopalynology by Mattu (1989) in Himachal Pradesh revealed that predominant pollen sources were *Trifolium* sp., *Plectranthrus* sp., *Eucalyptus* sp., *Sapindus* sp., *Toona* sp., *Brassica* sp., *Fagopyrum* sp., *Impatiens* sp., *Berberis* sp., *Taraxacum* sp., *Syzygium* sp., *Citrus* sp., *Nephelium* sp., *Tilia* sp., *Pyrus* sp., *Prunus* sp., *Rubus* sp., *Robinia* sp., *Medicago* sp., *Melilotus* sp., *Prinsepia* sp., *Salix* sp., *Dalbergia* sp., *Adhatoda* sp., *Helianthus* sp., *Polygonum* sp., *Cassia* sp. and *Salvia* sp. The minor pollens in the honey samples were *Amaranthus* sp., *Ageratum* sp., *Vernonia* sp., *Sonchus* sp., *Solidago* sp., *Senecio* sp., *Chichorium* sp., *Potentilla* sp., *Bidens* sp., *Aster* sp. and members of the families Lamiaceae, Asteraceae, Papilionaceae, Poaceae, Convolvulaceae, Rhamnaceae and Rosaceae in 64 honey samples collected from 30 localities (Mattu *et al.* 1988). Agashe and Rangaswamy (1997) conducted melissopalynological analysis of five *T. iridipennis* honey samples from Karnataka and found that *Psidium guajava* and *Syzygium cumini* were the predominant pollen types while all the five samples were unifloral. Roubik and Moreno (2009) reported 72 botanical species belonging to 65 genera in 41 families from the different honey of Meliponinae. Agashe and Rangaswamy (1997) conducted melissopalynological studies on different honey samples and pollen loads of *T. iridipennis* from Western ghat area of Karnataka and identified *Sapindus indica, Milletia ovalifloia, Syzygium caryophllatum, S cumini, S. jambolan, Mallotus phillippensis, Tabubia arjentia, T. avalenedae, Sclerophylum pentandrum, Hopea whitiyana, Kidia calicina, Ailanthus malabarica* and *Ceiba pentandra* as pollen sources.

Floral calendar for stingless bees (*Tetragonula* spp. and *Lepidotrigona* spp.)

A floral calendar for beekeeping is a calculated timetable that indicates the approximate date and duration on the flowering periods of the important nectar and pollen plants in a given area or region. The floral calendar is one of the most useful tool for apicultural extension worker. It enable him to inform the beekeepers on what to expect from availability of different bee forage so that they can manage their colonies in the most rational manner. The experienced beekeeper will have acquired much of this information by himself over the years, but published charts are also available in many areas. Above all, beekeeping in any specific area cannot be developed to its capacity without a proper understanding of the calendars for the different foraging zones (Smith 1960).

Chauhan and Singh (2018) developed a floral cycle for stingless bees in the lower hills in Nagaland. They described the flora season-wise to enable the stingless beekeepers manipulate the meliponiary as per the floral requirement of bees. A total of 89 plants were identified and categorised on the basis of their availability throughout the year. These include floricultural, vegetable, horticultural, medicinal and weed plants. Similarly, 64 plants were identified in foothills of Nagaland (Singh *et al.* 2016). Karupasamy and Jayaraj (2016)

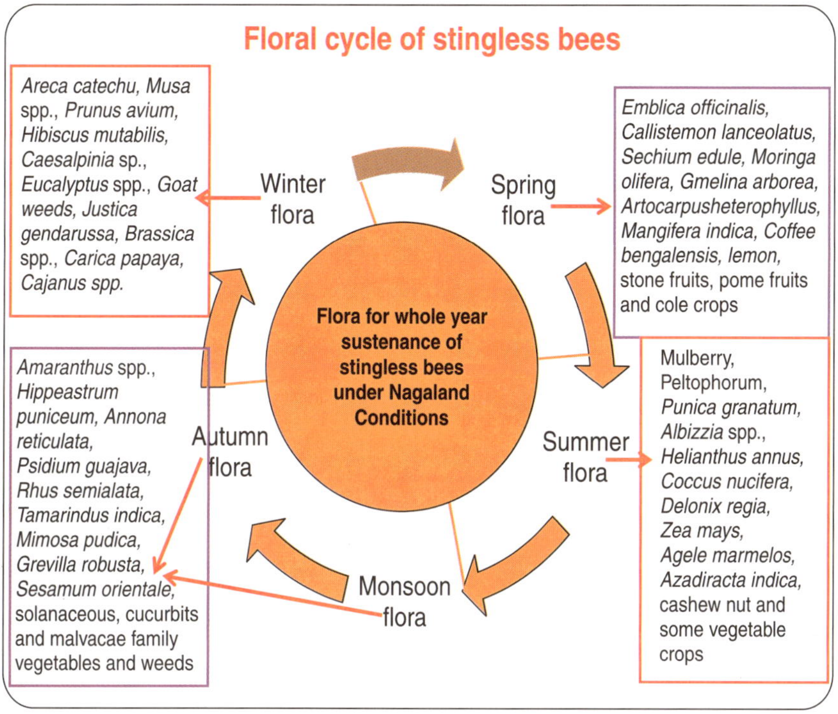

Fig 8.1. Floral cycle of stingless bees at Medziphema area of Dimapur, Nagaland (drawn by Chauhan 2018).

also identified 45 plant taxa belonging to 29 families and non-floral sources utilized by *Tetragonula iridipennis* for pollen, nectar and resin. The foragers of *T. iridipennis* were also found to collect non-floral resources like fruit juice, fruits kept for sale in the market and from fallen or damaged mango and jasmine fruits. The families Arecaceae and Fabaceae had a significant importance as the bees collected pollen mostly from the plants of these families. Similarly, floral calendars for bees are prepared in different states of the country to provide the reliable suggestions to the beekeepers for seasonal manipulations.

Thus, for stingless bees, different researchers have developed floral cycle throughout the globe for better understanding of bee's necessities and ensure proper management practices for better production of hive products and pollination management. In dearth periods which may vary for different regions, scientific management using floral calender, ensures artificial feeding of bee colonies for their sustenance and harvest optimum honey and other hive products during honey flow period. With the establishment of floral cycle, Meliponiculture is now expanding and is being adopted by different farmers as an alternative source of income generation.

■

CHAPTER 9

Management of Stingless Bees

Stingless bees are important group of bees which sustain the ecological processes of the planet. Many different species of stingless bees are reared in various parts of the globe. In the western countries, several tribes are associated with these bees. Similarly, in India, especially in north eastern and southern states, several tribes are in deep association with these bees. They practice the stingless bees culture from time immemorial in their myriad traditional styles or methods. With the advancement of rearing and management techniques, the stingless beekeeping is now expanding and adopted by new farmers and entrepreneurs in its scientific mode. The different managements adopted by meliponiculturists paved a new way to look at these important bees with new perspectives. The different strategies are practised and evolved for management of stingless bees throughout the planet. It is a fact that stingless bees are a vital resource to assure plant biodiversity in many natural ecosystems and more so, fulfill the growing agricultural demand for pollination. However, stingless beekeeping remains an essentially informal activity across developing countries where technical knowledge is scarce and management practices lack standardization (Jaffe *et al.* 2015).

Fig 9.1. Traditional beekeeper rearing two different stingless bee species in a single log hive.

Traditional management

Stingless bees are of great significance in terms of social, economic and religious aspects in different regions

Fig 9.2. a, Stingless bee hives placed around the kitchen for rearing; b, Traditional beekeeping with stingless bees in Nagaland, India.

of the globe where the people developed numerous systems of management and breeding of these insects. In tropical America, from México to Brazil, this activity goes back to the Pre-Hispanic era (Quezeda *et al.* 2001; Kent 1984; Crane 1992; Cortopassi-Laurino *et al.* 2006; Cortopassi-Laurino and Nougiera 2016). The age-old traditional knowledge and management practices associated with stingless bees still exist among indigenous communities in Mexico and Latin America. In Mexico, there are four areas where indigenous inhabitants are rearing stingless bees (Hendrichs 1941; Dixon 1987). Similarly in India, around 18 tribes and 9 sub tribes are associated with traditional stingless bee keeping in Nagaland (Singh 2016).

In all these areas, people use important management strategies and practices for the extraction of derivative products and breeding of new colonies. Meliponiculture, an important part of mesoamericans and tribals of India has gradually developed over the centuries with the sole purpose to harvest the products produced by stingless bee, which is a significant representation of nutritional and medicinal dietary component. The honey is used mainly as a medicinal supplement and treatment of ailments such as ocular infection, fractures, muscle pain, sprains, cutaneous wounds as well as gastrointestinal and respiratory illness (NBHM 2018; Vit *et al.* 2004).

Mesoamericans call pollen as "*pasacuareta*" in the Balsas region, and simple pollen in Nagamese which is consumed either alone as raw or mixed with honey by the Nagas, Khasi, and other tribal regions for respiratory infections and "weakness" or fatigue (NBHM 2018; Singh 2016). The wax which is an important hive product has been a valuable trade resource since colonial times as the Naga people use wax to exchange goods and even animals with people of Myanmar region.

However today, these many traditional practices associated with meliponiculture are found conserved only in few specific areas within the boundaries of Mexican tropics and Eastern India. In historical times, the cerumen derived from stingless bee colonies as a hive product was used to insulate and decorate the weapons for hunting and killing animals and even during the era of tribal wars (NBHM 2018). Similarly, in other regions of western and southwestern Mexico, the extraction of honey was basically from the species *Scaptotrigona hellwegeri* Friese, 1900. Historically, this was an activity of economic and socio-political importance as it was used as an offering to the Valle de Mexico which was the seat of the Mexican Empire (Dixon 1987).

The primitive extraction methods in the Balsas region was a kind of guarded activity that systematically sought specific nests and was considered a specialized trade but that ultimately resulted in the destruction of the nests (Gonzalez *et al.* 2016).

Technical advancements in management

Since time immemorial, Meliponiculture was practised only by the tribal people and researchers were relucant on honeybee research then. Today, the practice of meliponiculture still persists despite the ignorant ages and gaps. However, there has been resurgence in interest for this particular activity which has created impetus for strategies in order to rescue traditional meliponiculture making it more efficient through scientific studies with the goal of meliponini conservation and alternative productive projects (Gonzalez *et al.* 2016). This recent resurgence is directly linked with the growing demand for natural, organic, and homeopathic products which include honey, pollen, propolis and bees wax. All along the years, new techniques and production methods have been developed (largely through academic institutions) that has been termed "integrative meliponiculture." This has allowed for more efficient and sustainable management of these insects wheresoever they are present (Gonzalez *et al.* 2016).

Integrative meliponiculture is a modern practice that considers not only the production factor but also the conservation and maintenance of viable stingless bee colonies as well as the integration of the families or lineage of those interested in participating in this activity and also, the implication for community benefit (Gonzalez *et al.* 2016; Chauhan and Singh 2019). On the production front, the objective of integrative meliponiculture is to procure goods like honey, pollen, and propolis from a particular nest while encouraging specialized ecosystem benefits like pollination in greenhouses and agricultural fields as well as the landscape in general. Environmental education is also an important aspect of this activity wherever possible. Integrative meliponiculture limits the destruction of wild nests and favours their maintenance, rescue, and propagation. This activity also helps to limit the traffic and introduction of non-native species in the regions of stingless bee's natural distribution. These

Fig 9.3. Beekeeper inspecting scientific stingless bee hive of *L. ventralis* in the Meliponiary.

principles are the main challenge for meliponculture currently developed in various regions of the planet (Gonzalez *et al.* 2016).

The developement of new scientific hives for different species (Heard 1988: Sommijier 1999; Lindemann and Arya 2002; Punith 2014) increased the percentage of successful establishment of bee colonies with minimum loss to the brood during extraction of hive products (Chauhan and Singh, 2019). The bee colonies can also be manipulated for pollination due to more portability.

Multiplication of stingless bee colonies

For successful Meliponiculture, the beekeepers are required to multiply the colonies taking the seasonal management into consideration. According to Dixon (1981), beekeepers can acquire new colonies either by collecting natural nests from forests, by capturing new swarms using trap-nests, by purchasing them from other beekeepers or by multiplying their own colonies. Colony multiplication, whereby one colony is split into two or more colonies, is considered the most sustainable approach to obtain new colonies, since it involves low economic costs (a new box) and does not require the removal of feral bees from their native habitat (Cartopassi- Laureno *et al.* 2006). Colony multiplication is an integral part of stingless bee management because more colonies will ensure production of more honey and also, sell colonies to other beekeepers. The beekeepers who have received trainings on scientific meliponiculture are more likely to multiply colonies more successfully than those who have not, highlighting the importance of acquiring the necessary management skills to multiply colonies. This is because multiplying colonies of some species is not easy, requiring precise knowledge on the best times to divide and also the ability to recognize and carefully extract the brood discs at the appropriate developmental stage (Gonzalez *et al.* 2016). Moreover, newly multiplied colonies are usually weak and vulnerable to infestation by pest habitat (Cartopassi- Laureno *et al.* 2006).

Queen rearing

The methodical rearing of stingless bees in hives, termed as meliponiculture, is a highly sustainable activity because beekeepers can work at home near native forests as well as help to conserve endemic bee species to acquire bee products (pollen, honey, propolis) and promote effective agricultural pollination (Cortopassi-Laurino *et al.* 2006; Slaa *et al.* 2006; Contrera *et al.* 2011). Stingless bees are highly social bees and exhibit a division of labour for the welfare of the whole colony. The bee colony consists of queen, worker and males. The worker is responsible for all the chores of hive except egg laying while males do not share any work in development of nest except mating, whereas queen is the mother of the whole colony and thus command a key important role in the colony. She lays fertilized and unfertilized eggs depending upon the necessity of the colony. In some species, when there is no queen for a long duration, the ovaries of some of the workers start developing while in most other species, the colony slowly collapsed and died. As such, queen rearing is also an important part in scientific management of stingless bees.

Honeybees (Apini) and stingless bees (Meliponini) which are highly eusocial bees have evolved special adaptations that enable the colony to overcome queenlessness. Stingless bees breed and groom new queens naturally throughout the year and two different modes of queen rearing exist among them. The first is the rearing of gynes of similar size to that of workers in worker cells, which is seen in *Melipona* (Velthuis and Sommeijer 1991). The second is the rearing of larger gynes in specially-built royal cells which occur in all stingless bees except *Melipona* (Engels and Imperatriz-Fonseca 1990; Imperatriz-Fonseca and Zucchi 1995; Imperatriz- Fonseca *et al.* 1997; Velthuis and Sommeijer 1991; Sakagami 1982).

The process of queen rearing has three kinds of bionomic constraints in stingless bees:

1) Mass-provisioning of brood cells require all the nutrients necessary to support ontogeny of immature stages are deposited at once, before the queen's oviposition.
2) Brood cells are sealed right after the queen oviposits.
3) There is a strong socio-behavioural disorganization usually occurs when a colony is deprived of the queen's influence (Lacerda and Zucchi 1999). For these reasons, natural requeening after queen loss which usually occurs in honeybees, has been regarded as highly improbable in stingless bees.

Several research workers from various countries tried queen rearing with different species of stingless bees, viz. *Frieseomelitta varia* (Lepeletier), *Scaptotrigona depilis* (Moure) and *S. postica* (Latreille) (e.g. Hartfelder and Engels 1992; Baptistella *et al.* 2012; Menezes *et al.* 2013). A new method of stingless bee queen rearing was developed for *Plebia droryana* by Santos *et al.* (2016) and for its introduction and acceptance in the daughter or queen less colonies. The larvae fed with 66 μL of food developed into queens which were having acceptability rate of 75%. The method of queen rearing developed is described below as per Santos *et al.* (2015, 2016).

(a) In the apiary, strongly populated *P. droryana* colonies with newly built brood combs containing only eggs and larval food were chosen. These combs were decapped to allow extraction of the larval food. Other older brood combs, i.e. 1–3 days old were used as a source of newly hatched larvae for grafting. The age of the larvae was estimated by observing their size and position in the comb whereby the younger larvae were smaller and located adjacent to brood cells which contained only eggs.
(b) In the laboratory, approximately 66 μL of previously homogenized larval food was transferred to each cavity on acrylic plates with 50 cavities (each 5.8 mm diameter × 7.5 mm deep, i.e. 198 mm^3) using a manual pipette (10–100 μL).
(c) Three rearing assays were performed (trial 1: n =50; trial 2: n =50; trial 3: n =100 larvae) by transferring *P. droryana* larvae and larval food to the

cavities, being careful to retain the original orientation of the larvae seen in the brood cell, i.e. with the spiracles in contact with the air (if this is not carefully arranged the larva can die by asphyxiation). The larvae were moved using a wooden stick tipped with an entomological pin modified into a spoon shape.

(d) The queen-rearing plates were placed into hermetic plastic containers (7×11×17 cm) and kept inside an incubator at 25 °C in complete darkness (0L:24D).

(e) The relative humidity inside the plastic container was maintained between 98 and 65 % during the whole period of larval development.

After 35 days of larvae transplant, the queen emerged and were found healthy as natural queens. These queens were then given to the queenless colonies having newly emerged worker bees and more than 75 % acceptability was recorded.

Faustino *et al.* (2002) successfully reared the *Trigona (Frieseomelitta) varia* queens in Brazil and the artificially reared queens started laying eggs after 19 days of transplant into colonies.

In some species of stingless bees like *Tetragonula carbonaria*, the workers cannot lay eggs and thus slowly the whole colony collapses which is a constraint for beekeepers. It therefore, becomes important to remain careful while dividing or multiplying the colonies so that each daughter colony have at least one queen cell for their successful establishment. Nunes *et al.* (2014) found that in Australian stingless bee, *T. carbonaria*, two worker cells combined to form queen cell when colony becomes queenless. In India, several workers tried to rear stingless bee, *T. iridipennis* queens in different parts of Karnataka, Tamil Nadu and Nagaland for multiplication of colonies or to provide queens to queenless colonies for sustainability of stingless bee keepers (Biennial report AICRP HB&P 2019).

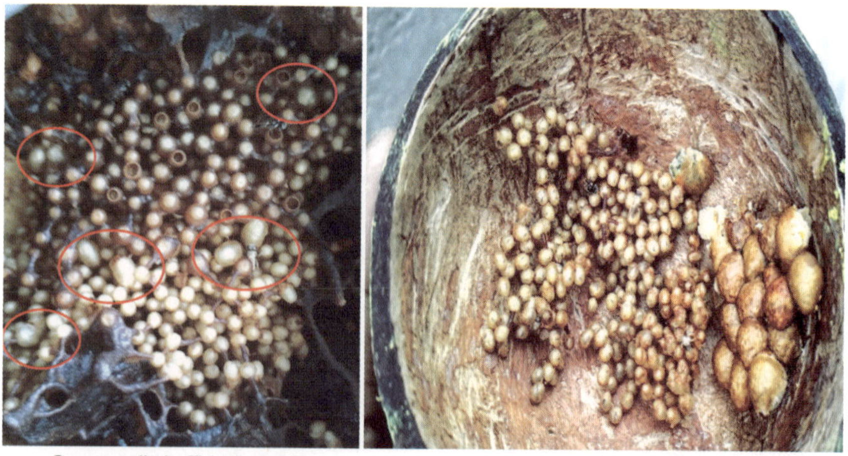

Queen cells in *T. iridipennis* (pc Sarvanan)

New colony with transplanted queen cell

Fig 9.4. Queen cell rearing and division of stingless bee colony at Tamil Nadu, India.

Swarming

The foundation of new colonies in highly eusocial bees (honeybees and stingless bees), occurs through swarming. It is natural division of colony in which some bees (may be half or more) leave the colony along with new queen and this young swarm settles generally in the nearby area of the colony. The swarming or natural division of colonies in stingless bees is very interesting. Some of the scout workers search a suitable place in the vicinity of the mother hive. The workers finalise the new place and start cleaning, transportation of food reserves and hive material to the new nest. They build the colony entrance, confined to some species only and construction of food pots and some workers start to forage while others remain visiting the mother nest. When the new nest is ready, more workers fly together with a virgin queen which will perform the nuptial flight soon after arriving at the new nest. The contact between mother and daughter nest can last from few days to several months in accordance to availability of resources and the species of stingless bees (Nogueira-Neto 1954; Juliani 1967; Terada 1972; Wille and Orozco 1975; Inoue *et al.* 1984; Engels and Imperatriz-Fonseca 1990; van Veen and Sommeijer 2000).

Period of swarming

The multiplication of stingless bees occur by swarming. When a colony has reached a certain size and a new nest place is found, some worker bees start transporting building materials to the discovered place. More and more bees will fly to the new nest over the next few days and at the end, a queen from the old nest will migrate to the new nest and begin producing eggs there. Thereafter, a new colony will be established, and soon, the flying visits between the two colonies will cease later (http://www.fao.org/3/i0842e/i0842e07.pdf).

The proportion of workers present in the daughter colony was about 30 percent when the daughter colony became virtually independent from the mother colony (i.e. about one week after swarming). The mass flight of males continued for 2 days after swarming around the daughter nest. The young queen remained heavily surrounded by workers and made repeated buccal contacts with them for about 5 days after swarming. Oviposition also started one week later. In most of the species, mating between a new queen and drones usually takes place outside the nest (Inoue *et al.* 1984). It occurs when the queen has reached her peak of brood rearing activity under the stimulus of incoming pollen and nectar, mainly in the late spring or early summers or in the late autumn or onset of winter season depending upon the floral conditions of the area. This generally occurs during the period before honey flow (Chauhan and Singh 2020).

Cause of swarming: The main cause for the swarming in honeybees is described by Chauhan and Singh (2020) which are as follows:

Overcrowding and lack of ventilation: When the bee colonies becomes populous due to the small nest size, the colonies starts developing queen cells which matures to form new queens.

Presence of old queen: the communication and control among bees mainly depend on the pheromones produced by the queens. When the queen is old or the colony is over expanded, the workers receive less pheromone and the instinct of division is developed.

Sudden honey flow: when there is sudden flow of nectar in the flora, the colony starts collecting it and at one stage, the instinct of division starts which leads to swarming.

Lack of space for egg laying and honey storage: When the nest is small and brood rearing is more due to incoming food sources, the colony starts preparing for swarming.

Problems due to swarming

Swarming is the natural division of colony. Inoue *et al.* (1984) found that approximately 30% of workers go with the daughter colony. Due to this, there is loss of working force or foragers. This further resulted in a change in behaviour of the remaining working force. They start focusing more on rearing of new brood cells and thus the morale of colony is not favourable for honey collection (Chauhan and Singh 2020).

Indication of swarming

In many species, many bees do not go to field for foraging and thereby, create additional crowding that result in clustering of bees outside the hive. In some species, the bees behave normally but may be found to engage themselves in transportation of hive materials to outside the colony.

Time of swarming

In stingless bees, the swarming is gradual. The workers transport hive materials and food to the new location slowly and when nest is ready, only then the new queen leaves with other workers. The higher activity of swarming can be seen from 1000 h to 1500 h.

Prevention and control of swarming

Swarming is a natural process of multiplication of colony but it also has some negative effects on the mother colony because it ultimately leads to weakening of mother colony. A single colony may issue one to several swarms depending upon the condition of colony. To prevent swarming:

- Divide the colony when swarming instinct is observed.
- Avoid overcrowding in the hive.
- Remove the queen cells at regular interval if you are sure about the mother queen.
- Provide shade and ventilation to the colonies.
- By removing old queen followed by introduction of queen cell. In some species (*T. iridipennis*), a numbers of queen cells are present while in others (*Tetragonula carbonaria*) queen cells are not present.

If swarming is not controlled in stingless bees, trap nests can be placed in the meliponiary. In Brazil, trap nest method is the only legal means of collecting stingless bee colonies from the field (Conama 2004). The technique of trap nest was studied by different researchers to establish the effectiveness of trap nests for stingless bees (Inoue *et al.* 1993; Malkowski *et al.* 2006; Alvarenga 2008). Oleveira *et al.* (2012) also evaluated the effectiveness of trap nests for domiciliation by swarms of different stingless species found in Brazil.

Fighting swarms

In stingless bee species, the workers aggregate in front of the hives and after some hours of aggregation, hundreds of bees can be found dead on the ground. This aggregation of the bees to attack other bees is termed as fighting swarms. It is shown that the Australian stingless bee *Trigona carbonaria* sometimes display a striking collective behaviour, known as a 'fighting swarm' in which thousands of workers fight and die (Gloag *et al.* 2008). The reason behind the fighting of swarms is described by various researchers (Schwarz 1948; Michener 1974).

Gruter *et al.* (2016) reviewed the fighting swarms in stingless bees. According to them, the nest is the all-important centre of the bee's activity in providing food and shelter for the entire colony together with the necessary architecture for rearing subsequent generations of workers and reproductive individuals (Roubik 1983, 1989, 2006). Finding and acquiring the resources to maintain the colony is often a time consuming and laborious activity. A honeybee forager, for example, may visit several hundred flowers during a single foraging trip to fill its crop (Ribbands 1949).

Stealing the stored resources of another colony therefore has substantial benefits in terms of saving time and energy, and we would expect that collective attacks evolve when the overall payoff for both individual and colony is higher than from independent acquisition (Iyengar 2008). Since these resources (e.g. honey or pollen) have similar chemosensory features (e.g. sugars and floral volatiles) as flowers, a switch from foraging to stealing might not have required a major change in sensory adaptations (Breed *et al.* 2012). This behaviour, also called cleptobiosis (Breed *et al.* 2012) or cleptoparasitism (Iyengar 2008), is known to occur in several stingless bee species, with the nature of these intercolony attacks varying considerably in terms of mortality (from none to thousands) and duration (hours to weeks).

Attacks occur in two different contexts, foraging (raiding of supplies) and swarming (usurpation of the entire nest to establish a new colony). When attacks eventuate in hive usurpation, the attacking colony not only gains the entire provisions of the hive (e.g. supplies of food, wax and resin), but also the nest cavity itself. Thus, the attacking colony does not loose a substantial amount of its resources as a result of swarming, which could be particularly advantageous when competition for food is intense (Quezada-Euan and Gonzalez-Acereto 2002).

Usurpation might also be favoured in an environment where nest-sites are scarce (Foitzik and Heinze 1998; Rangel *et al.* 2010). However, attempts to gain resources by force can have considerable mortality costs if attacked

Fig. 9.5. *Tetragonula iridipennis* workers aggregated outside log hive during swarming.

Fig. 9.6. *Tetragonula iridipennis* workers fighting during swarming in hives located on trees.

colonies mount a strong defence (Cunningham *et al.* 2014; Johnson 1987) and invaders risk acquiring diseases and pathogens from the resident colony (Breed *et al.* 2012; Lindstrom *et al.* 2008).

Stingless bees have diverse species having different foraging preferences. The genera *Lestrimelitta* (21 species) (Gonzalez and Griswold 2012) and Cleptotrigona (1 species) (Eardley 2004) are obligate cleptoparasites with no flower foraging behaviour, but rather stealing resources by invading other stingless bee (and occasionally honeybee) nests (Michener 1946; Muller 1874; Roubik 1989; Sakagami and Laroca 1963). Workers of these robber species are equipped with powerful mandibles that allow them to kill workers of defending colonies with ease (Nogueira-Neto 1970). Other stingless bee species, such as *Melipona fuliginosa*, are known to raid nests but are not exclusive robbers (Nogueira-Neto, 1970): this facultative raiding may occur under particular environmental conditions, such as food shortages (Downs and Ratnieks 2000) or may represent a transitional state from facultative to obligatory robbing (Nogueira-Neto 1970). In *Tetragonula iridipennis* (Chauhan and Singh 2020 data not published) the fighting behaviour was observed in Nagaland (Fig 9.5; Fig 9.6). They found that these fightings often end up leaving the colonies weak or sometimes mass death of robbed colony.

Artificial colony multiplication/ dividing of colonies

The bee colonies can be divided into daughter colonies by several methods such as parallel method with gynes of colony division, parallel method with queen cells, budding methods etc. The best season for increasing number of colonies is in early spring. It is done by dividing the colonies which are not as strong as others and sparing these colonies from honey production. Such colonies can be divided into other daughter hives by taking a portion of brood along with honey and pollen cells. A new queen or queen cells should be given to the new colony. At the entrance, put cerumen material from the mother colony so that a portion of the old workers also enters the new hive.

Uniting of bee colonies

In honeybees, different techniques are available to unite the weak colonies before honey flow period but in stingless bees, lot of research is required to establish the validity of the methods. The stingless bee keepers in the country first identify the colonies which are weak or queenless. The brood and food sources are transferred from very weak colony to the other weak colony and if queen is not present, queen cell is given. If food resources are less, the bees are artificially fed with sugar syrup. Similarly pollen stores are made available from the corbicular pollen collected by honeybee colonies with pollen trap. The united colonies should be checked after one week for any development or requirement. Sometimes, we can also strengthen weaker colonies with workers or brood from stronger ones, provide supplementary feeding if required, or detect whether a colony is queen-less (Chauhan and Singh 2020).

Precaution: Keep the better queen and remove poor queen before uniting. Check for the status of pests and any fungal infection before uniting the colonies.

When the necessity arises?

- During fall, uniting weak colonies which cannot overwinter well in hilly regions.
- Just prior to honey flow, uniting weak colonies to make one strong colony.

Absconding: It is a phenomenon desertion in which the whole stingless bee colony leaves the hive.

Causes

- Shortage of food reserves
- Attack of bee enemies/ diseases
- Too much disturbance in handling

Prevention

- Keep colonies strong and ensure that each colony has good reserves of food.
- Avoid broodlessness in the colonies and provide new queen or queen cell if required.
- Manage colonies accordingly and check the colonies for diseases and attack from bee enemies.

Feeding of bees

Bees require supplementary feeding when the food stores are poor. Supplementing the food stores of bee colonies with sugar syrup and pollen substitute/pollen supplement is known as supplementary feeding. Several stingless bee keepers provide bee collected pollen and sugar syrup to bee colonies when flora is scarce outside.

When do the bees need feeding?

- When the colonies do not have sufficient food stores for undergoing wintering.
- When new colonies are being formed and transferred or divided.
- During drought and even before honey flow.

What is the best food? : Honey is the best source of food for bees. To substitute honey stores, sugar in different forms (as syrup, dry or candy) can be fed to the needy colonies.

Feeding methods

Honeybees collect surplus food in the form of nectar and pollen when bee forage is available in plenty. Nectar is converted into honey and stored for future use when there comes dearth of period. Similarly, pollen is processed to form bee bread for later use to meet protein requirement of the colony. But with commercial interests in mind different hive products are harvested from bee colonies which need to be supplemented for their survival. Colonies are fed sugar in different forms to supplement honey and pollen substitute or supplement in place of pollen. Therefore, it is important to know processes for preparation of different types of feeds given to bee colonies as well as methods of feeding.

CHAPTER 10

Biotic Stresses in Meliponiculture

Stingless bees are one of the smallest honey producing bees. Hundreds of species are distributed from subtropical to tropical regions of the planet. In India, *Tetragonula iridipennis* Smith and *Lepidotrigona* spp. are the main native stingless bee species available. These common stingless bees are small to medium sized with vestigial sting. They are also called "dammer bees" as they collect resins for constructing their nest along with wax produced from dorsal plates of their body. *Tetragonula* refers to their triangular shaped abdomen and *iridipennis* refers to their iridescent wings. *Tetragonula* (Trigona) *iridipennis* was first described from Ceylon by Smith in 1954. The most common species of stingless bee found in India was known by name *Melipona iridipennis* Smith. Later, this species *Melipona* was found to be restricted only to neotropics and as such, was transferred to the genus *Trigona* (Michener 1974). It was distributed only in India and Sri Lanka (Sakagami 1978).

The stingless bees live in permanent colonies, nesting in old walls, logs, crevices and naturally concealed places. Stingless bee honey is popular for its antioxidant and antibiotic properties (Chauhan and Singh 2018) and hence, fetches higher price in the market. The defence behaviour of these bees is by chasing and biting the intruders. They become entangled in the hairs and get enter into the nose, ears, and eyes of the intruders.

Social bees are host to a variety of parasitic organisms like entomopathogenic fungi, mites, nematodes, parasitoids, protozoa and viruses. These stresses may adversely affect their survival and reproductive process (Hempel 1998). Honeybees are known to have been attacked randomly by several enemies, viz. wax moth, varroa and tropilaelaps mites, lizards, ants, spiders, toads and birds and diseases like Thai sac brood, Stone brood, Nosema, American foul brood, European foul brood and many others (Chauhan and Singh 2020; Divya *et al.* 2019). Similarly, stingless bees have biotic stresses which hinder their development or sometimes leads to total loss of colonies resulted in less honey productivity. The different pests of stingless bees are phorid flies, Megachilids, ants, birds, syrphid flies, pollen beetles, mites, lizards, frogs/toads, predatory wasps, spiders (Hart *et al.* 2018). However, a few bacterial diseases are also recorded from brood of stingless bees (Shank *et al.* 2017).

Shanks (2015) in Australia worked on stingless bees. He disclosed that similar to honeybee colonies, stingless bee colonies can also be affected by predating or scavenging pests, however stingless bees are unaffected by *Varrosis* and to date no viruses have been detected. Pests of importance include

Syrphid fly, *Ceriana ornata australis* Saunders 1845 (Diptera, syrphidae), phorid fly, *Dohrniphora trigonae* Disney 1995 (Diptera, Phoridae), scuttle fly, *Megaselia scalaris* Loew 1866 (Diptera, Phoridae), mantis flies, *Plega hagenella* Westwood 1867) (Neuroptera: Mantispidae) (Maia-Silva *et al.* 2013), sand wasp *Bembix musca* Handlirsch 1894 (Hymenoptera, Crabronidae), braconid wasp *Syntretus trigonaphagus* Gloag, Shaw & Burwell 2009 (Hymenoptera, Braconidae), grain itch mite *Pyemotes tritici* LaGrèze-Fossat and Montagné 1851 (Acari, Pyemotidae) (Menezes *et al.* 2009) and also small hive beetle *Aethina tumida*. In contrast to honeybees, there is almost no documentation of brood disease in stingless bees. In fact, the only credible report is that of Kerr who observed possible diseased pupae in *Melipona quadrifasciata* and *M. bicolor bicolor* (Kerr 1948; Nogueira-Neto 1997). Jayalekshmi (2015) studied the pests and diseases of stingless bees.

Pests of stingless bees

Flies – *Hermetia illucens*

A dipteran fly *Hermetia illucens* L. which thrive on decaying organic matter has been recorded as a pest of stingless bee, *T. iridipennis* in Kerala. Both the maggots and adults were observed from within the stingless bee hives. It was discovered that the larvae of dipteran fly, *H. illucens* voraciously fed on the pollen, honey, brood and cerumen deposits of *T. iridipennis* (Nisha 2002). The adult fly is smoky and black winged, easily recognized by the presence of two translucent spots on first abdominal segment while the maggots were plumpy, firm, leathery with tiny yellowish black head and dull white body colour. Simoes *et al.* (1980) described *Mealonctia sinistia* Borgmeier as an endoparasitic phorid fly of *Nannatrigona (Scaptotrigona) postica* Latreille in Brazil. Sommeijer (1999) reported phorid fly (*Pseudohypocerrspo* sp.) as a major pest of stingless bee in America, which fed on pollen and brood. The maggots and adults of the phorid fly *Megaselia scalaris* Loew affected the nested colonies of all the five stingless bee species in Kakamega forest, Kenya (Kiatoko 2012).

Nitidulid beetles – *Aethina* sp.

These beetles are one of the important destructive pest of stingless bee colonies, causing damage to brood cells, pollen and honey pots. The adult beetle shows dark brown body about 5 mm in length. The adults are usually broad, flattened and oval in shape. The antennae are clubbed and elytra are shortened in these beetles. In Kenya, the brood of *Meliponula bocandei* Cockerell, *Meliponula ferruginea* Cockerell (reddish brown), *Meliponula ferruginea* Cockerell (black), *Meliponula lendliana* Friese and *Hypotrigona gribodoi* Magretti stingless bee species are vulnerable to invasion by *A. tumida* grubs (Kiatoko 2012). He also reported other coleopteran pests like *Rhizoplatys mucronatus* Beauvois and *Tenebroides mauritanicus* L. from the pollen pots of feral colonies of *M. bocandei* and *M. ferruginea* from Kakamega forests, Kenya. *Carpophilus* spp. and *Brachypeplus* spp. are hive beetles and sap beetles found to attack stingless bee colonies (Spooner *et al.* 2018).

Syrphid flies

One of the most common and severe pests of stingless bee colonies is the syrphid fly, *Ceriana ornata australis* Macquar. *Ceriana ornata* is 12 mm long, with bright orange-yellow and black markings and is frequently observed hovering near nests during summer. This fly has been observed in all regions where stingless bee colonies are present and is seldom observed in areas where stingless bees do not occur (Klumpp 2007).

Megachilids – *Megachile* spp.

Megachilid bees are also an important pest of stingless bees. The adults of the megachilids steals the cerumen from the entrances of the *Tetragonula* spp. While, in *Lepidotrigona* spp., megachilids are not reported (Chauhan 2018). *Megachile disjuncta* was reported as a pest of stingless bees in Tamil Nadu and *M. lanatus* in Nagaland (Chauhan and Singh 2020).

Bugs – Assassin and reduvid bugs

The predators of stingless bees can be spotted near the bee nests. Assasin bugs and reduvid bugs are generally found feeding on the adults of stingless bees. According to Jayalekshmi (2015), reduvid bug, *Acanthaspis siva* adults are morphologically black coloured with elongated head, distinct narrowed neck and prominent segmented legs. Long stylets are also present. The nymphs camouflag themselves by attaching bits of bee cadaver onto their body. They remain near the entrance tube of the colonies. While, in assassin bug *Sycanus* sp., the adult insects are orange or black in colour. The abdomen is curved upward with a large black spot and is fully covered by wings at rest. The stylets are short, twice as thick as their antennae and curved. The forelegs are raptorial. The adults and nymphs of *Sycanus* are observed around the hives, crevices and on tree trunks in the vicinity of stingless bee hives. The eggs are brown in colour and are laid in clusters and cemented to each other. Nymphs are wingless and reddish orange in colour. The females of the assassin bug *Apiomerus pillipes* Santschi are predators of the genus *Melipona* (Silva and Santana 2004) in Brazil. Wattena and Jongjitvimol (2007) reported that *Pahabengkakia piliceps* Miller as a predator of the stingless bee, *Trigona collina* (Smith) in Thailand. *Acanthaspis siva* Distant (Reduviidae: Hemiptera) was reported as a predator of stingless bees and honeybees in Kerala (Premila *et al.* 2013).

Mites

A number of mites are found in the nests of stingless bees. Some mite species feed on debris while others feed on bee products including brood. Flechtmann *et al.* (1974) found a small whitish mite, *Neotydeolus therapeutikos* (Flechtmann and de Camargo) in the nests of *Trigona postica* (Latreille) in Brazil. *Meliponopus palpifer*, a phoretic deutonymph (hypophus) was noticed in *Melipona seminigra* (Merrillae) in Brazil (Fain and Fletchmann 1985). *Neocypholaelaps phooni* Baker a new species of mite was found in nests of stingless bees in Malaysia as reported by Baker and Delfinado (1985). Baker and Baker (1988) observed two mites belonging to the genus *Eumellipites* in the nests of *Tetragonula* (*Trigona*). A new species of mite *Neocypholaelaps malayensis* (Pilsbry) was found in the nests of *Trigona itama* (Cockerell) and *T. iridipennis* in Malaysia as described by Delfinado- Baker *et al.* (1989). Smiley *et al.* (1996) reported a new species of mite *Hypoaspis* (Mesostigmata: Laelapidae), as a pest of stingless bee *T. iridipennis*. Nisha (2002) reported a mite *Amblyseius* in bamboo hives of stingless bees in Kerala. She further reported that mites feed on the pollen lead to the destruction of the entire colony. Another new species of mite, *Carpoglyphus lactis* L. was reported by Vijayakumar *et al.* (2013) from the nests of *T. iridipennis* in Tamil Nadu. *C. lactis* mites were found to infest the pollen store initially and subsequently spread over the brood cells. Menezes *et al.* (2009) reported the infestation of stingless bee nests by the mite *Pyemotes tritici*, which killed four colonies of *Tetragonisca angustula* and one colony of *Frieseomelitta varia* in Brazil.

Spiders

Different species of spiders were recorded as a predator of stingless bee colonies by various researchers (Spooner *et al.* 2018; Jayalekshmi 2015). Spiders are major predators of stingless bees. Eight different predatory spiders observed by Jayalekshmi (2015) were *Thomisus projectus, Thomisus lobosus, Neoscona nautica, Menemerus bivittatus, Plexippus paykulli, Hersilia savignyi, Heteropoda venatoria* and *Crossopriza lyoni*. Two pathogens, viz. *Paecilomyces variotii* and *Aspergillus niger* were also observed. Craig *et al.* (1996) reported

Fig 10.1. Spider trying to prey on stingless bees.

that *Nephila clavipes* L. a large orb spinning spider feeds on *Trigona fluviventris* (Friese) by making large yellow webs. *T. fluviventris* is attracted to yellow colour and hence they are trapped in the webs of *N. clavipes* more frequently than in any other spider web. Wattena and Jongjitvimol (2007) reported that *Nephila maculate* F. a golden orb web spider and *Argiope* sp. a golden garden spider were predators of *Trigona collina* (Smith) nests in Thailand. In Kerala too, different species of spiders (*Thomisus* spp.) were found to predate on stingless bees (Devanesan *et al.* 2009).

Ants

They are common predators of bees in tropical and subtropical region of Asia. They attack the hives in mass and take out virtually everything, viz. honey, dead or alive adult bees and brood. Ants belonging to subfamilies of Dorylinae and Ecitoninae, which included the army ants, were fearsome predators of honeybees and stingless bees (Hamida 1999). Raakhee (2000) observed *Solenopsis geminata* Wheeler (ants) entering the weak colonies of stingless bee hives to rob pollen stock from colonies in different locations of Kerala. Ants (*S. geminata*) were seen entering the hive especially in initial stages when the colonies were just divided (Devanesan *et al.* 2009). The ant *Myrmicaria* sp. as a predator of stingless bee species in Kenya was reported by Kiatoko (2012). The weak colonies of stingless bees are more susceptible to ant attacks. Weaver ant, *Oecophylla smaragdina* was reported as a predator of stingless bee adults in Tamil Nadu (AICRP 2019).

Lizards, frogs and toads

The minor pests of stingless bees include lizards, frogs and toads. In Australia, *T. carbonaria* colonies are attacked by frogs and toads (Spooner *et al.* 2018). In India, especially in the north eastern region, *T. iridipennis* and *Lepidotrigona* spp. are attacked by lizards frequently.

Birds

Birds cathches the adult stingless bees either from the nest entrances or while foraging on flowers. The common house sparrow is identified as a minor predator of stingless bees in India. The African blue fly catcher bird, *Elminia longicauda teresita* Antinori was reported as a major hunter of forager stingless bees, *M. ferruginea* (black) and *M. ferruginea* (reddish brown) by Kiatoko (2012) in Kenya.

Other pests

The stingless bee *M. lendliana* was one of the predators that frequently robbed propolis from *M. bocandei* hives which were put in place to seal gaps in their hives (Kiatoko 2012). He also reported that the honeybee *A. mellifera* robbed the honey stored in pots in the nests of the five stingless bees at the Kakamega forest in Kenya. In India, honeybees were found to attack the stingless bee's entrance for propolis (Chauhan and Singh 2020).

Mammals

Large animals like civets (*Paradoxurus hermaphrodites* Pallas), bears (*Melursus ursinus* Shaw), honey badgers (*Mellivora capensis* Storr), tamandus (*Tamandua tetradactyla* L.), tayras (*Eira barbara* L.) and chimpanzees (*Pan troglodytes* Blumenbach) were also reported as important enemies of stingless bees (Roubik 1995).

Natural tactics to avoid pests

Entrance closing in stingless bees: In stingless bees, it is observed that when stingless bees are constantly exposed to pest attack or robbers, they close their entrance for weeks to months (Fig 10.2). During this time, the workers do not leave the nest and instead, stay inside (Hart *et al.* 2018; Chauhan and Singh 2020). The colony becomes weak as the queens does not lay eggs and the adults feed on reserved food only.

Stingless bees are capable to control the little infestations of pests like small hive beetle, *Aethina tumida* by capturing the adult beetles with the cerumen (Fig 10.3) and the larvae were killed by the worker bees (Halcroft *et al.* 2011).

Diseases in stingless bees

Stingless bee brood is not much susceptible to diseases unlike honeybees. Some researchers have reported few bacterial and fungal diseases in stingless bees (Jayalekshmi 2015).

Fig 10.2. Stingless bee, *T. iridipennis* colony attacked by honeybees and closed hive entrance of bee colony (*Courtesy:* Robert, Spooner Hart and Avinash Chauhan).

Fig 10.3. Trapped beetle adult in cerumen and larvae killed by workers (*Courtesy:* Helcroft *et al.* 2011).

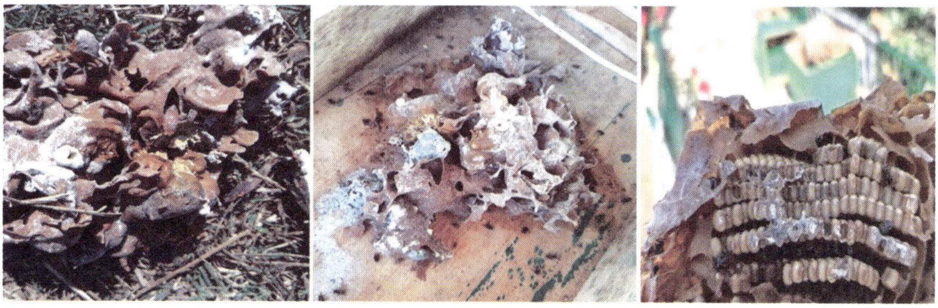

Fig 10.4. Fungi infected hive cells and brood cells of stingless bee.

Hermetia illucens

Sycanus sp.

Acanthaspis siva

Sycanus sp. attacking *T. iridipennis*

Worker bee trapped in web spider

Lizard outside entrance

Stingless bee attacking robber *A. cerana*

Fig 10.5. Different pests of stingless bees (*Source:* Jayalekshmi 2015; Spooner 2018; Chauhan and Singh 2020).

Shanks *et al.* (2017) observed a bacterial disease in stingless bee, *T. carbonaria*. They isolated a disease-causing bacterium, *Lysinibacillus sphaericus* (Firmicutes, Bacillaceae), from worker and queen larvae, brood cell provisions and honey stores. Pathogenicity experiments confirmed this bacterium as the causal organism. Incubation period of the disease was 22 days. This is the first confirmed record of a brood pathogen in stingless bees. The name of the disease was coined as "Shanks brood disease".

Not much information is available on stingless bee diseases except some references from Kerala. Jayalekshmi (2015) reported few fungal strains from abandoned hives, pollen pots, honey pots, propolis and from the dead bees lying inside the hive floor. Two fungal isolates of *Paecilomyces variotii* (Bainier) was isolated from Thiruvananthapuram district while a single isolate of *Aspergillus niger* (Tiegh) was isolated from Pathanamthitta district. Incidence and cultural characters of isolates from stingless bee colonies were recorded.

■

CHAPTER 11

Bee Poisoning

Stingless bees are highly developed social insects that provide an ecologically crucial important service of pollination resulting into diversity and conservation of flora and fauna (Chauhan *et al.* 2019; Witter and Tirelli 2014). Bees of the genus *Tetragonula, Trigona, Melipona, Scaptotrigona* and *Tetragonisca* can be used to pollinate field and greenhouse crops due to their important characters discussed in other chapters also traits like fidelity, docility, absence of functional sting and adaptation to artificial domains (Slaa *et al.* 2003, 2006; Barth *et al.* 2008; Giannini *et al.* 2015a; McGregor 1976; Heard 1988, 1999; Nates Parra 2005). There are a considerable number of articles that deal with the toxicology and the effect of insecticides on bees but little literature is published on stingless bees (Costa *et al.* 2015; Jacob *et al.* 2013; Lourenco *et al.* 2005; Moraes *et al.* 2000; Soares *et al.* 2015; Tome *et al.* 2015; Valdovinos-Nunez *et al.* 2009). The harmful effects may be caused due to the direct exposure of bees to toxic pesticides or through indirect contact with their residues. Direct exposure occurs either from treatment of bee hives for disinfestation purpose or bees visiting the fields at the time of spray and the indirect exposure occurs from the spray drift or bee foraging on sprayed crops. There are multiple means of exposure to pesticides that affect bees. Being in direct contact with a pesticide solution, powder from treated seeds or the application of pesticides in granular form to the soil are the most common ones. Oral exposure to residues include products translocated by the plants that are present in the pollen or nectar of plant or associated plants within or at the crops border and nearby crops, all the exposed aerial applications, the application of granules to the soil, the seed treatment or the absorption of residues from previous grown crops are considered highly hazardous (EFSA European Food Safety Authority 2013c; Godfray *et al.* 2014). In addition, it is assumed that toxicity response also depends on intrinsic factors, such as ecology, biology, physiology, morphology and extrinsic factors such as climate and flora.

Pesticides, particularly neonicotinoid (neurotoxic) insecticides, were earlier considered as the main culprits behind the large decline of pollinators although the common impression of their importance in agriculture remain as such, despite controversies that have been generated by the absence of supporting field evidence studied of their risk analysis (Cutler *et al.* 2014; Henry *et al.* 2015; Lawrence *et al.* 2016). The significant attention on a particular group of pesticides namely the neonicotinoid insecticides diverts the focus of all earlier studies which explored the effects of other pesticides and chemicals utilized in agriculture (i.e. agrochemicals). However, this trend is slowly shifting, and have began to recognize the potential risk inflicted by other pesticides, including fungicides, herbicides, biopesticides (i.e. pesticides of natural origin), formulation "inerts", fertilizers as well as different mixtures of various chemicals on stingless bees (Johnson 2015; Mullin *et al.* 2015; Rodrigues *et al.* 2016).

Poisoning symptoms in bees

Honeybees react differently to different pesticides, and most herbicides and fungicides are less toxic to bees than are insecticides. The most common symptom of bee poisoning is the sudden appearance of excessive number of dead bees in front of the hives. The following figures have been established as guidelines for assessing the extent of pesticide poisoning: 100 dead bees per day is the colony's normal death rate; 200-400 dead bees indicate a low level of pesticide poisoning; 500-1,000 dead bees indicate a medium level of pesticide poisoning; over 1,000 dead bees indicate a high level of pesticide poisoning. Another common symptom is the lack of foraging bees and aggressiveness in bees may also be caused by most pesticides. Chlorinated hydrocarbons and organophosphorus insecticides commonly causes stupefaction, paralysis and abnormal activities of bees (Johansen *et al.* 2010).

The exposure to fipronil in *Scaptotrigona xanthotricha*, a species of stingless bees, it initially induced a highly excited condition, represented by wing movement, vibration and circular movements at all once. Later it was observed that slowness in movement, tremor and a displacement difficulty occured which was also experienced in another stingless bee, *M. scutellaris* when it was exposed to the same compound orally and topically (Lourenco *et al.* 2012). The symptoms in *Tetragonisca angustula* showed trembling of individuals till their death. The individuals showed full spread of wings and legs before death after less than 4 h of the application of pesticides. With thiamethoxam poisoning, individuals of both species (*M. scutellaris and Tetragonisca angustula*) started to move rapidly throughout the box; minutes later they slowed down and were paralyzed, agreeing with the symptoms observed for *S. postica* when exposed to imidacloprid (Soares *et al.* 2015); in some cases the bees died clinging to the mesh of the experiment boxes.

The dead individuals had swollen abdomen with contracted legs and wings. In India, the impact of Neonicotenoids on pollinators (*Apis mellifera*, *Apis cerana* and stingless bee, *Tetragonula iridipennis*) revealed that the foraging of bees, brood rearing, incoming and outgoing activity and food stores are highly affected for the first 7- 12 days of exposure while after second week of exposure the hives started recovering. In *Tetragonula iridipennis*, the exposiure to pesticides resulted into destruction of queen cells by the workers. In weak colonies, the death rate was higher as compared to strong bee colonies (Anonymous 2019). Some common symptoms for initial identification of bee poisoning are given below.

1. Presence of dead bees at the entrance of hive or colonies.
2. Presence of dead bees among the frames or on the bottom board.

3. Inability to recognize guard bees.
4. The degree of aggressiveness in bees can also be a vital factor.
5. Visual observation of infighting among nest mates.
6. Presence of semi-paralysed bees crawling around or nearby the hive.
7. Appearance of immediate decline in food and brood.
8. Presence of dead as well as deserted brood in the hive.
9. Deficient ability in recognition of nectar and pollen.
10. Observation of a decreasing population in a colony.
11. Obvious contamination present in bee products.

Causes of bee poisoning

The majority of bee poisoning occurs when insecticides are sprayed on crops during the blooming period. Insecticidal dusts are attached gradually to foraging bees and ultimately mixed with the pollen on their hind legs. It is a fact that Penncap-M and Sevin are especially dangerous because they may be stored together with pollen and later kill newly hatched workers during the next season (Mayer *et al.* 1999). Atkins *et al.* (1981) observed that it depends on the toxicity level of insecticides on bees, wherein insecticides with LD_{50} less than 2 μg/bee (2,000 ng/bee) are highly toxic.

The toxicity of pesticides are also based on the mode of action and entry into the body of insects. In the case of thiamethoxam, stingless bee, *S. xanthotricha* have revealed a higher tolerance to topical exposure than that of *T. angustula*, even though there is a high toxicity to oral exposure. The different causes of exposure to insecticides are:

- Application of pesticides on blooming crops.
- Drifting of toxic chemicals onto different parts of flowers and gradually, pollen and nectar.
- Feeding the bees with food and water from impure sources.
- Application of broad-spectrum insecticides such as chlorinated hydrocarbons, synthetic pyrethroids etc.
- The type of formulation applied for insecticides such as dusts, EC etc. which are more harmful than WP and granules.
- Factors such as types of spray, fineness of spray, stage of crop, weather condition and age of the colony.
- The use of insect growth regulators may also inhibit brood production.
- Application of herbicides indirectly affect bees by damaging the foliage.
- Using diesel oil as a carrier during insecticide formulations.

In the stingless bee species *Melipona quadrifasciata* the lethal and sub lethal effects of the neonicotinoid imidacloprid and the bio-insecticide spinosad, indicated high susceptibility to insecticide doses when administered orally, with LD_{50} of 23.54 and 12.07 ng a.i./bee for imidacloprid and spinosad respectively. Imidacloprid also impaired the respiration system of workers and overall group activity and flight while spinosad significantly impaired only the flight of workers in spite of showing higher oral toxicity to adult workers than imidacloprid. These simultaneous findings indicate the hazardous nature of both imidacloprid and spinosad upon adult workers of the native pollinator *M. quadrifasciata*. Therefore, bio insecticides can also pose great risk due to their lethal and sub lethal components.

The potential negative consequences of pesticides to native stingless bees are equally important for the management (Barbosa *et al.* 2015; Casida *et al.* 2011). The insecticides chlorpyriphos and phosmet are potentially dangerous (Table 11.1 and Table 11.2) to *S. bipunctata* and *T. fiebrigi* both topically and by ingestion (Dorneles *et al.* 2017). Jacob *et al.* (2013) had reported a topical LD_{50}: 0.54 ng/bee and oral LC_{50}: 0.24 ng μL-1 of fipronil in the usual diet of *S. postica*. Meanwhile, Lourenco *et al.* (2012) who worked on *Melipona scutellaris*, recorded a topical LD_{50} of 0.41 ng/bee and an oral LC_{50} of 0,011 ng μL^{-1} fipronil diet.

Table 11.1. Acute oral toxicity of insecticides (commercial formulations) to *Scaptotrigona bipunctata* and *Tetragonisca fiebrigi* workers

Insecticide	Species	LC_{50} (95% CI) µg a.i./µL diet
Chlorpyriphos	*S. bipunctata*	0.0112 (0.0095–0.0130)
	T. fiebrigi	0.0018 (0.0017–0.0020)
Phosmet	*S. bipunctata*	0.0245 (0.0206–0.0285)
	T. fiebrigi	0.0236 (0.0201–0.0271)

Table 11.2. Acute topical toxicity of insecticides to *Scaptotrigona bipunctata* and *Tetragonisca fiebrigi* workers

Insecticide	Species	LC_{50} (95% CI) µg a.i./µL diet
Chlorpyriphos	*S. bipunctata*	0.0110 (0.0092–0.0128)
	T. fiebrigi	0.0033 (0.0030–0.0037)
Phosmet	*S. bipunctata*	0.0087 (0.0075–0.0100)
	T. fiebrigi	0.0083 (0.0072–0.0095)

Among pesticides, nicotinoid pesticides are the most toxic, in the following descending order by permethrin, diazinon, and methomyl. Insecticides other than neurotoxic compounds may also cause high acute mortality to stingless bees. The activity of azadirachtin, one of these compounds that primarily acts as an insect growth regulator, was detected in the stingless bee, *Melipona quadrifasciata* (Barbosa *et al.* 2015b), and was also reported from the bumble bee, *Bombus terrestris* (Barbosa *et al.* 2015c). However, in both instances these insecticides represented a lesser threat than did the neonicotinoids imidacloprid and spinosad (Barbosa *et al.* 2015b, 2015c). Understanding the high acute toxicity of leaf fertilizers is a greater challenge. Toxicity was observed for oral exposure to field-relevant concentration of copper sulfate in adult worker bees of *Friesella schrottkyi*, a small stingless bee species common in southern and southeastern Brazil (Rodrigues *et al.* 2016). The high lethal toxicity may be caused by copper, a heavy metal that can bind to vital cellular components (e.g. proteins, enzymes) and compromise cell function while generating reactive oxygen (Halliwell and Gutteridget 1984; Phillips and Hilliker 1990; Cox 1999).

Effects of poisoning in stingless bees

The nutritional behavior of stingless bees also increases the probability of larval exposure to contaminated food unlike honeybees, which are progressive feeding bees. Stingless bees practice mass provisioning, i.e. before egg laying provide the brood cells with sufficient for the full larval development stages (Sakagami 1982). Likewise, if stingless bee foragers collect contaminated pollen or nectar, the larvae will be provided with contaminated agrochemical diet throughout the different stages of post-embryonic development. Furthermore, the larval food of stingless bees consists of a mixture of pollen, nectar and glandular secretions, with great amounts of proteins and less water tin comparison to the royal jelly of honeybees and as such, there is a great probability of larval contamination in stingless bees (Hartfelder and Engels 1989; Tome *et al.* 2012; Lima *et al.* 2013). The stingless bee, *Tetragonula iridipennis*, on exposure to Imidacloprid, Dimethoate and Thiamethoxam have shown less foraging activity, brood rearing and damage of queen cells.

Management of bee poisoning

In the management of bee poisoning, the basic principle is to avoid the exposure of bees to toxic effects. This could be achieved with the help of both bee keepers and the farmers. The practices to be followed by bee keepers include the following:

1. Bee colonies should be maintained where use and drift of pesticide is minimum.
2. Intimate co-operation with farmers to avoid irrational use of pesticides.
3. Feeding of colonies with sugar syrup at the time of pesticide application to reduce bee foraging.
4. Use of pesticides based on requirement only.
5. Informing the bee keepers in advance about the spray programme.
6. Use of less hazardous, selective and repellent (methyl salicylate, R-874) insecticides.
7. Spraying in the evening when the bee activity subsides.
8. Granules, EC are preferred compared to dusts.
9. Avoid formulations with attractants like Sevimol during crop bloom period.
10. Development of bee strain resistant to toxic effects of pesticides.
11. Addition of adjuvant Sylgard 309 silicone surfactant to reduce the bee mortality.

It is well known fact that no insecticide is safe to bees but can be relatively safe with respect to each other. Stingless bees are highly susceptible to all neonicotenoids, organochlorines, organophosphates, biopesticides and physical pesticides. So, while shifting the stingless bee colonies for pollination under open field and protected conditions, the pesticides should not be sprayed or sprayed atleast 3 weeeks before flowering. A list of comparatively safer insecticides is given below:

Less hazardous insecticides

1. **Granules**: Fenthion, phorate, aldicarb and lindane
2. **EC**: Endosulphan, phosalone, fluvalinate, menazon and lindane

Highly hazardous insecticides

1. **Dust**: Carbaryl, diazinon and fenvalerate
2. **WP**: Carbaryl
3. **EC:** Chlorpyriphos, cypermethrin, deltamethrin, diazinin, dichlorvos, dimethoate, ethion, fenitrothion, fenthion and fenvalerate.
4. **SL**: Imidacloprid

CHAPTER 12

Hive Products of Stingless Bees

The stingless bee exists in almost every continent of the globe. The honey produced by this bee has been widely used across time and space since time immemorial. The distinctive feature of this honey is that it is stored naturally in the pot (cerumen), thus contributing to its beneficial properties, especially in the wound healing process (Jalil *et al.* 2017). The rearing of stingless bees for hive products and pollination management is termed as Meliponiculture and the place where the stingless bee hives are kept for rearing is called Meliponiary. Besides honey, stingless bees provide bee wax, cerumen and pollen (Fig 12.1) which are of high medicinal value and thus fetches better prices than honey of other bees (*Apis* spp.) (NBHM 2018).

Honey

Honey is the prime hive product derived from stingless bees. The honey is collected from various blooms as nectar and is carried to the nest for feeding the brood with pollen. The particular honey is unique as it is derived from the rich vegetation in native environments. It has a distinctive sweetness mixed with a sour and acidic taste. The appearance of stingless bee honey ranges from light amber to dark. In contrast to the stingless bee population around the globe, the availability of their honey is lower as compared to that of the common *Apis* honeybees. This is due to the limited knowledge about this honey, which has resulted in being less popular in terms of its industrial production, shelf-life, and quality standard (Guerrini *et al.* 2009; Rodriguez *et al.* 2012). Even though there are studies on the potential of honey in treating several health problems but the studies are mostly focussed on Tualang and Manuka honey when compared to the stingless bee honey due to its low production of honey (Roowi *et al.* 2012). Stingless bee honey can act as anti-inflammatory (Borsato *et al.* 2014), anti-cancer (Kustiawan *et al.* 2014; Yazan *et al.* 2016), anti-microbial (Zainol *et al.* 2013; Nobre da Cruz *et al.* 2014; Nishio *et al.* 2016; Medeiros *et al.* 2016; De Sousa *et al.* 2016; Nishio *et al.* 2016) and possessed antioxidant properties (Duarte *et al.* 2012; Almeida da Silva *et al.* 2013; De Sousa *et al.* 2016). However, many benefits of stingless bee honey has been abandoned in modern medicine due to the paucity of systematic scientific studies for supporting its medical properties (Perez *et al.* 2006). The composition of stingless bee honey differ from other species according to some physicochemical parameters (Ozbalci *et al.* 2013) and other studies prove that honey from stingless bees are more valuable and it has been used for a long time to treat various diseases (Souza *et al.* 2006).

Yaacob *et al.* (2017) compiled the information on medicinal properties of stingless bees discovered by various researchers (Tables 12.1, 12.2 and 12.3).

The price of stingless bee honey, compared to the honey from *Apis* spp. is relatively high at around US $40 per liter in Brazil (Cortopassi-Laurino *et al.* 2006) US $80 in Malaysia (Basrawi *et al.* 2017) and US $45 in Thailand. In India, the price of stingless bee honey ranges from 2,000-4,000 ₹/kg ($25 - $50/kg) varying with season and region (Chauhan and Singh 2020).

Propolis

Another beneficial product from the stingless bee is "propolis," geopropolis, or cerumen. Propolis is hive product containing chiefly of bee's wax and resin derived from plant tissues or exudates (Bancova 2009). On the other hand, cerumen is a mixture like propolis but with the addition of the mandibular secretion of the stingless bee during its construction (Santos *et al.* 2009; Simone and Spivak 2010). The functions of these hive products are slightly different where the cerumen is used as honey storage pot and to mummify trespassers as well as to ensure that the environment in the hive is

Table 12.1. Antioxidant analysis of stingless bee honey (Yaacob *et al.* 2017)

Authors and year	Types	Method	Outcomes
De Sousa *et al.* (2016)	*M. subnitida* Ducke and *M. scutellaris* Latrelle	ABTS cation radical scavenging and DPPH scavenging method	Honey of *M. subnitida* showed the higher antioxidant activity compared to *M. scutellaris* in both assays.
Almeida da Silva *et al.* (2013)	*Melipona* (*Michmelia*) *seminigra* Merrillae)	ABTS cation radical scavenging	The samples showed the highest antioxidant capacity has a higher total phenolic content.
Duarte *et al.* (2012)	*Plebeia* spp.	Ferric reducing antioxidant power (FRAP) assay and (DPPH) scavenging assay	Honey from *Plebeia* spp. has higher content of total phenolics, flavonoids and antioxidants capacity compared to *Apis* sp.

Table 12.2. Anti-inflammatory analysis of stingless bee honey (Yaacob *et al.* 2017)

Authors and year	Study population	Types	Method	Outcomes
Borsato *et al.* (2014)	Animals. Male Swiss mice	Stingless bee from species *Melipona marginata*	Mice induced with 12-O-tetradecanoylphorbol-13-acetate-induced for ear edema model.	The honey extract (1.0 mg/ear) had the ability to reduce ear edema with an inhibitory effect of 54 ± 5%.

Table 12.3. Cytotoxicity analysis of stingless bee honey (Yaacob *et al.* 2017)

Authors and year	Study population	Types	Method	Outcomes
Yazan *et al.* (2016)	Sprague Dawley rats aged 5 weeks (n= 24)	Honey from *Trigona* sp.	Rats were injected with azoxymethane (15 mg/kg) and the treatment groups were given via oral administration of *Kelulut* honey (1,183 mg/kg body weight) twice per day for 8 weeks.	The total number of aberrant crypt foci (ACF) and aberrant crypts (AC) and crypt multiplicity were significantly reduced. This suggests that stingless bee honey has chemopreventive properties.
Kustiawan *et al.* (2014)	Human cancer cell lines HepG2 SW620 ChaGo- 1 KATO- III BT474	*Trigona incisa, Trigona apicalis, T. fuscobalteata, T. fuscibisca*	Crude extract was screened for *in vitro* cytotoxicity against the cell lines using the 3-(4, 5-dimethylthiazol- 2-yl) - 2, 5-diphenyltetrazolium bromide assay.	Crude extract of stingless bee has cytotoxic effect to HepG2 cell line.

sterile, whereas propolis is used as an internal insulating layer and for sealing the extra space surrounding the hexagon-shaped nest combs. Consequently, due to the storage of honey in the cerumen pots, the quality of the stingless bee honey is influenced by the infiltration of phytochemicals from the cerumen (Temaru *et al.* 2007; Massaro *et al.* 2011). In the cerumen pots, the nectars undergo 3 different processes of transformation before transformed into honey. The first transformation that takes place is a physical change wherein a large part of water usually evaporates from the stored nectar. It is followed by a biological transformation, which is mainly a fermentation process by yeast and bacteria. These microorganisms which originate from a suitable microenvironment chosen by the bee, play a symbiotic role with the colony (Menezes *et al.* 2013). Finally, a chemical transformation materialise when the worker bees secrete enzymes from their cephalic glands that are able to hydrolyse the sucrose from the nectar into fructose and glucose (Menezes *et al.* 2013; Anderson *et al.* 2011). To some extent, it is difficult to refuse that the characteristics of honey simply vary depending on the bee species, nectar sources and climate, the noteworthy similarity in stingless bee honey is the water content that is commonly higher than in other types of honey (Souza *et al.* 2006; Bijlsma *et al.* 2006). As a result, the rich water content invites the microorganism to inhabit and thrive inside the honey. Most of these microorganisms are probiotics with their beneficial enzyme secretion that works along with the enzymes from the stingless bees to enrich and conserve the quality of the honey (Menezes *et al.* 2013; Costa and Da-Cruz 2005).

Bee wax

Bee wax is an important hive product extracted from the hives. The wax is produced by workers from the glands present on dorsal side of the abdomen unlike honeybees who produce it from glands present in ventral abdominal side (Dreyling 1903; Cruz-Landim 1967). In *A. mellifera* and stingless bees the wax glands become active after 6 to 18 days of emergence (Hebling *et al.* 1964).

In the most famous medicine book of China, "The Shen Nong Book of Herbs", bees wax was recognised as a top medicinal ingredient. It is praised for its beneficial influence on blood and energy systems and the overall balance of the body (Behrer 1999). Several researchers found stingless bee wax is different from honeybees in physical as well as chemical composition. Wax of stingless bees was used in South America for making of candles (Saint-Hilare 1830). In Java, the wax of *Tetragonula iridipennis*, is used by the natives to produce Battik, a substance used to colour textiles (Schultz 1907). Similarly, Southern Brazil natives were using the wax for attaching the feather finery along with other articles (Peckolt 1893). Stingless bee wax was used in Australia for decorating their hairs and to protect the paintings from rain (Rayment 1935). Herbert (1945) disclosed the importance of stingless bee wax from different continents of the globe. He found that, the indigenous people of different continents were using the stingless bee wax since time immemorial for different works. NBHM (2018) described the use of bee wax from Naga inhabited regions in India. Milborrow *et al.* (1987) described the wax of *T. australis* as colourless but that the brown coloured nest material is derived from the inclusion of pollen masses (*Eucalyptus* sp.) and solid material from the inner wall of the nesting tree. The solid residue comprised 12 and 30% by weight. An analysis of the nest material of *T. australis* by gas chromatography/chemical ionization-mass spectrometry have shown stingless bee wax comprised a hydrocarbon fraction (90%), esters (6%), and free acids (4%). The main saturated hydrocarbons present in wax were C_{27}, C_{31}, C_{33}, C_{25} and C_{29} where the latter two were less abundant while C_{23} and C_{35} were present in small amounts. Hydrocarbons (C_{31}, C_{33} and C_{35}) which contain one double bond were found and traces of the diene C_{35} were also detected. The compounds of ester fraction were not identical with those in bees wax of the honeybee *Apis mellifera* and moreover, the acid fractions were also quite different. The wax of *T. australis* contained saturated free acids: C_{10} (trace), C_{12}, C_{14}, C_{16}, C_{15} and C_{20}, monoenoic and dienoic C_{18} and C_{20} and traces of the trienoic C_{18}. The study on wax composition of two stingless bee species, *Trigona buyssoni* and *T. atomaria* by Blomquist *et al.* (2010) revealed that the wax of *Trigona (Trigonisca) buyssoni* and *T. atomaria* is much simplier in composition than that of *Apis mellifera* and consists of hydrocarbon (59.3 ± 3.5% and 70.9 ± 6.3%, respectively by weight), wax monoester (26.8 ± 3.5% and 25.5 ± 6.3%), free fatty acid (5.4 ± 1.6% and 1.7 ± 0.9%), primary alcohol (7.0 ± 2.4 and 1.0 ± 0.9%), and unidentified more polar material (1.5 ± 0.1% and 0.3 ± 0.5%). In the wax of *Trigona*, there was no detection of diols, hydroxy acids or their esters. The hydrocarbons in both species ranged in chain length from C_{21} to C_{37} and consist of *n*-alkanes (90.2% and 87.1%), alkenes (8.4% and 12.4%) and methyl branched alkanes (1.4% and 0%). The wax esters contain primary alcohol moieties of C_{16} to C_{30} in which the major components include tetracosanol, hexacosanol and octacosanol. The fatty acid portion of the wax esters have chain lengths of 12 to 30+ carbons with 18:1 comprising over 30% of this fraction. The majority of free primary alcohols range from C_{24} to C_{30} in both species with C_{24}, C_{26} and C_{28} components predominating.

Cavalkante *et al.* (2000) described the differences in development of wax glands in four stingless bee species that produce different quantities of wax. During the time in which maximum production of wax is expected, the

Stingless bee honey | Pollen

Wax | Propolis/cerumen

Fig 12.1. Different hive products of stingless bees.

histological data and average height of the wax epithelium, are in confirmity with the rates of wax produced by the species. In *Lestrimelitta limao* (Smith 1863) which is a species with cleptobiotic habits and thereby, a frequent robber of wax from other colonies, the height of wax epithelium was the lowest among the studied species.

Pollen

Pollen is the male gamates of the angiosperms which are collected by bees while visiting flora. When bees forage on the flowers, they collect pollen grains and pack them in the corbicular legs. This pollen is mixed with the honey and fed to the brood. Pollen is the source of protein which is very essential for egg laying. If pollen reserves are not enough, the queen bee avoids egg laying. In the hive, when flowering is huge during spring season, the bees collect extra pollen and store in the pollen pots for later use. This pollen contains many metabolites which are of higher medicinal value.

Extraction of pollen

The beekeepers extract the pollen from stingless bee hives during those months when pollen collection by bees is more. The pollen is extracted using pollen traps or manually from the pollen pots of the hive. The pollen is dried and then stored at low temperature for its use as supplementary feeding during dearth periods. Pollen is a rich source of proteins, vitamins and minerals and provides all necessary nutrients to the bees (Ferreira and Absy 2013). Pollen may be packaged and used as food supplements and also be added to infant food. It is also used in many cosmetic preparations (Absy *et al.* 2018).

■

CHAPTER 13

Pollination - An Essential Service

Pollination being an important component of the agro-ecosystem directly ensures increase in food security and improvement of livelihoods. In simple terms, it is the transfer of pollen grains from the anther which is the male part of the flower to the stigma, which is on the female part thereby resulting in fertilization (Fig 13.1). It is an important process in the plant reproduction without which reproduction for offsprings will not be possible. It is a complicated process that arose from coordination and perfection over the ages as plants coevolved with animals where the animals played the part of pollinators or pollination agents. The plants and animals coexist in same habitats. The coevolution between plants and animals in consideration to pollination eventually gave rise to the development of pollination syndromes. Pollination syndrome is a process where specific pollination agents pollinate specific plants or flowers. However, it is not a rigid arrangement as there are polyphilic flowers which attract and are pollinated even by different types of pollination. Pollination syndromes are a symbiotic relationship between the plants and the agents or animals in which both are benefitted mutually. The agents acquire nutrients from the flowers, whereas the plants benefit in the form of successful and improved reproduction. The later is very important in crop production. The angiosperms in plants exhibit two types of pollination, viz. cross-pollination and self-pollination. In cross-pollination, pollen grains are artificially transferred by insects from the anther to the stigma of another plant. A successful cross-pollination is followed by fertilization which leads to the production of offspring with heterosis or hybrid vigour. On the other hand, self-pollination is the transfer of pollen grains from an anther to a stigma of the same flower or plant. This process results usually into inbreeding depression where undesirable recessive traits are expressed in the offspring. Self-pollination transpire in species where pollination agents are scarce and in closed flowers where pollination agents have restricted access to the sexual structures of the flower (Free 1993).

Pollination syndromes

A set of adaptive morphological features exhibited by a pollinator to forage on plant reward while simultaneously the pollinated plants evolve floral traits to display such reward is called pollination syndrome (Faegri and Vander 1966).Within pollination syndromes, plants and animals exploit each other for their own advantage. Animals exploit the relationship to guarantee their reproduction and survival. They forage on the flowers for carbohydrates

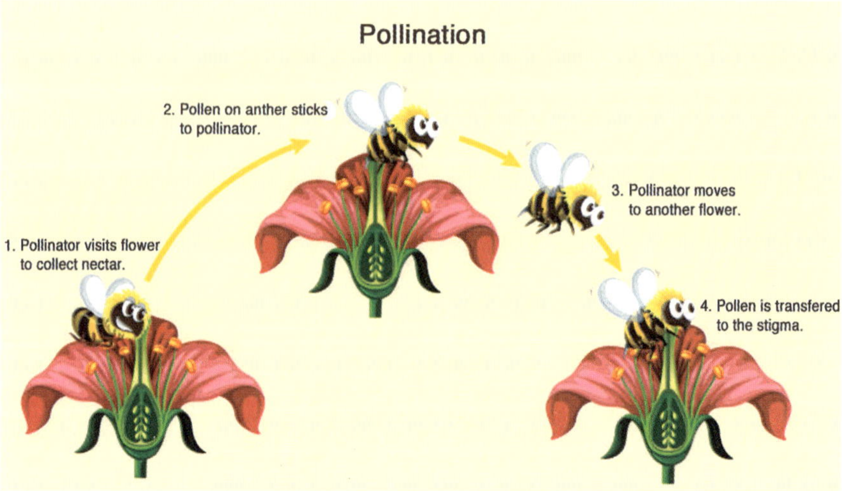

Fig 13.1. Role of bees in transferring pollen (pollination) (*Courtesy:* freepik).

(in nectar) and proteins (in pollen grains), which are the rewards. On the other hand, the plant guarantees itself with pollination and reproduction by providing the rewards and offering other attractive features like colour and scent (Roubik 1995).

Batra (1995) reviewed the role of bees and pollination requirements in the ever-changing environment. He compiled research sequels of different researchers where some of them stated that different pollinators possess different pollinating capabilities that vary with the bee species, viz. its size, hairiness, quickness, fidelity, longevity, learning ability, flight range, cold tolerance, season and flower handling ability etc. The relevant match between each bee species and its host plant is crucial, because a clumpsy adapted pollinator may even become involved as a parasite when they collect nectar and pollen without achieving profitable pollination (Allard 1911; Green 1955; Fronk and Slater 1956; Medler 1956).

There are different modes employed by the plants for the attraction of pollinators. Some produces good smell while some come up with attractive colours. However, scent plays a crucial role in cases where the pollination agents are deceived at the end. The deception involves mainly reproduction hues like pheromones and reproduction substrates or sites. Different plants produce flowers of different colours and the colours reflect light of different wave lengths. Since animals can perceive colour of specific wavelength, each type of animal will be attracted by a flower of different color. A combination of floral structure, reward, colour and scent from the plant and colour perception and nutrient preference by animal result in a specific pollination syndrome. The following pollination syndromes are recognized: cantharophily (beetles), melittophily (bees), myophily (by flies), psychophily (butterflies and moths), chiropterophily (bats), and ornithophily (birds). In these pollination syndromes, the pollen grains become attached on some parts of the animals and are eventually brushed onto the stigma. On the other hand, some pollen grains are transferred by abiotic factors (wind and water) from anther to stigma.

The different types of pollination modes on the basis of pollen carrying agents are described below:

Cantharophily

The type of pollination carried out by beetles are called Canthrarophily and the flowers in beetle pollinated plants are unspecialized. They are dull in colour, scented, and produce lots of pollen grains, the beetles feed on the pollen grains and other flower parts. The scent is emitted from different parts of the flower. In north-eastern conditions, blister beetle normally feed on pollen of *Hibiscus* spp. and is responsible for seed setting.

Chiropterophily

It is the mode of pollination accompanied by bats and flowers pollinated by bats are large, bowl-shaped, dark in colour, produce large amounts of nectar, pollen etc. and smell like rotten fruit. Bats pollinating the flowers are frugivores and nectarivores.

Melittophily

Bees forage on flowers to collect pollen and nectar and the pollination accomplished in this way is known as Mellitophily. The flowers are blue, pink, yellow in colour and some reflect UV light and produce lots of nectar and pollen grains.

Myophily

Flies are attracted by the mimicry of carrion and faecal matter, a phenomenon called sapromyiophily. In mango blooms, lots of dipterans can be observed.

Ornithophily

In this, the pollination is executed by birds and the flowers pollinated by birds are red in colour, tubular in shape, scented, and nectariferous (Medan). The birds are attracted to the red colour and scent while they obtain nectar as reward. Bird pollinators are mainly hummingbirds and sunbirds. These are hoverers which flip their wings while sucking the nectar. Others are perchers which sit on branches while sucking nectar.

Other pollination syndromes

In wind pollinated flowers (anemophily), the pollen grains are very fine and are released on dangling anthers. The stigmas are sticky and hang out of the flower like beard. This increases the chances of the pollen grains landing on the stigma. This type of pollen transfer occurs mainly in the grasses and some acacias. Water aided pollination occurs mainly in submerged water plants. Both water and wind pollinated flowers do not produce nectar and are not brightly coloured.

Floral resources and specialisation

Bees are highly social insects capable to feed on large number of angiosperms with unique specificity. According to Wappler *et al.* (2015) the evecative examples of insect pollination have stressed on closely related pollinator mutualisms such as figs and fig wasps (Ronsted *et al.* 2005) and yuccas and yucca moths (Pellmyr 2003). However, recent attention by pollination ecologists has focused on the broader spectra of pollinated plants by generalised pollinators such as bees. Bees have great impact on formulating hypotheses regarding specialization versus generalization in pollination mutualisms (Waser and Ollertron 2006, Ollerton *et al.* 2007).

Bees are pious herbivores that feed their larvae with mixtures of pollen and nectar or rarely, plant oils (Michener 2007). Robertson (1925) was one of the first to acknowledge that bees do not randomly collect pollen on flowers but that some species display floral specificity while harvesting only pollen on a limited number of plant taxa. He notably introduced the terms monolecty, oligolecty and polylecty to differentiate between pollen specialists and generalists accordingly. This classification however, incompetent to reflect the complex relationships between bees and their pollen hosts and was later updated by Cane and Sipes (2006) which was recently modified again (Table 13.1) by Muller and Kuhlmann (2008). It was learnt that even generalist bees depict a restricted range of pollen sources (Westrich 1989; Muller 1996; Muller and Kuhlmann 2008) wherein pollen may not be an easy manageable protein source.

The pollinators often exhibit generalized or specialized behaviour for pollen collection. The pollen collection from a great variety of pollen taxa by a foraging pollinator is referred to as polylecty (Canes and Sipes 2006). In contrast, specialization, in its pure form, is the collection of a single pollen type from a host-plant species and is known as monolecty (Canes and Sipes 2006). The vast transitional zone between these two end members is oligolecty, which ranges from broadly oligolectic pollinators that approach polylecty to narrowly oligolectic forms that merge into monolecty (Canes and Sipes 2006). The prevalence of these three major modes of pollen acquisition by foraging insect pollinators has been a contentious issue since the mid-1990s (Waser and Olletren 2006; Waser *et al.* 1996) and has currently created a central exploratory program in pollination ecology (Waser and Olletren 2006). Host specialisation has evolved multiple times independently in bees (Wcislo and Cane 1996). Oligolecty is frequent in solitary bees with short life cycles wherein the emergence of bees and host-plant flowering are tightly coupled (Hurd 1957; Danforth 1999; Minckley *et al.* 2000). Oligolecty is rare in eusocial bees. However, specialist bumblebees (*Bombus consobrinus* and *B. gerstaeckeri*) (Ponchau *et al.* 2006) are known to forage exclusively on pollen of *Aconitum* sp. Specialist bees are particularly common in some lineages such as Andrenidae, non-social Apinae, Lithurgini, Melittidae, Paracolletini and

Table 13.1. Classification of floral specificity of pollen collection by Robertson (1925), Cane and Sipes (2006), and an updated classification proposed by Muller and Kuhlmann (2008)

Robertson (1925)	Canes and Sipes (2006)	Muller and Kuhlmann (2008)
Monoloecty	Monoloecty Pollen collection from only one plant species even in the presence of one or more sympatric species of the same genus	Monoloecty Pollen collection from only one plant species even in the presence of one or more sympatric species under the same genus (see also Narrow oligolecty)
Oligolecty	Narrow oligolecty Pollen collection from two or morespecies belonging to one plant genus	Narrow oligolecty Pollen collection from two to several species belonging to one plant genus (pollen collection on only one plant species in the absence of co-flowering congenerics is referred to as a special case of narrow oligolecty)
	Oligolecty Pollen collection from two to several species that belong to one to four genera belonging to one family	Broad oligolecty Pollen collection from two to several genera belonging to one plant tribe, subfamily or family
	Eclectic oligolecty Pollen collection from two to four plant genera belonging to two or three plant families	Eclectic oligolecty Collection of pollen from two to four plant genera belonging to two or three plant families
Polylecty	Mesolecty Pollen collection from more than four plant genera belonging to two or three plant families or big tribes	Mesolecty Pollen collection from more than four plant genera belonging to two or three plant families
		Polylecty with strong preference Pollen collection from several plant families, but one plant clade (family, subfamily, tribe, genus or species) predominates
	Polylecty Pollen collection from various genera within four to < 25% of available plant resources	Polylecty s.s. Pollen collection from various genera belonging to at least four plant families
	Broad polylecty Pollen collection from various genera belonging to >25% of available plant families	

Rophitinae (Linsley 1958; Wcislo and Cane 1996; Michener 2000; Patiny *et al.* 2007; Larkin *et al.* 2008).

Foraging behaviour of stingless bees

Stingless bees (Apidae: Meliponini) are common visitors on flowering plants in the tropics but evidence about their importance and effectiveness as crop pollinators is still lacking for most plant species. They are known to visit the flowers of 90 crop species. They were confirmed to be effective and important pollinators of 9 species. They may make a contribution to the pollination of 60 other species but there is insufficient information to determine their overall effectiveness or importance. They have been recorded from another 20 crops but other evidence suggests that they do not have an important role because these plants are pollinated by other means. The strengths and limitations of stingless bees as crop pollinators were discussed by Heard (1999). The various aspects of their biology that impact on their potential for crop pollination are reviewed including generalized flower visiting behaviour of colonies, floral constancy of individual bees, flight range and the importance of natural vegetation for maintaining local populations (Heard 1999).

The foraging behaviour of bees is the criteria to assign them with the status of a pollinator. The foraging activities of bees can be assessed through direct observations of the flow of workers from their nests. Michez *et al.* (2008) found evidence that host-switches occur more frequently to morphologically similar, rather than closely-related host-plants suggesting that floral morphology plays a key role in host-plant evolution in bees. The record of quantity and quality of resources (pollen, nectar/water or water, resin, mud and garbage) that entered and/or left the nests in every 10 minutes during each hour from 0600 h to 1800 h is suggested for establishing the behaviour of pollinator (Bartelli *et al.* 2014). A single flower patch is often exploited by several bees as well as other bee species simultaneously (Boogert *et al.* 2006). Stingless bee prefers white or yellow flowers (Cortopassi-Laurino *et al.* 1991). They prefer small flowers (Wille *et al.* 1983) and dense inflorescences (Gumbert and Kunze 1990). Stingless bee stick to floral constancy, *i.e.* worker on a trip usually visits only one plant species (Ramalho *et al.* 1994). Ciar *et al.* (2013) concluded from their experiments that a height of 1 m from the ground was the most preferred for foraging by the bees. The feeders with 1 m height from the ground have the shortest distance from the bee hive at which the hive opening can easily be discovered by scouts/foragers. In regard to direction, they observed that bees did not follow consistent preferred routes or direction. The choice of feeder varies depending on current environmental conditions (such as wind speed and direction) or might be due to non-apparent behaviour (such as random search). In the foraging activities of *T. biroi*, the bees employed Area-Restricted Searching Strategy and Marginal Value Theorem. Bees have the tendency to maximize the time spent in an abundant floral area until food sources are depleted and thereafter, they choose other foraging sites which is often abundant in the nearest neighbour (Ciarl *et al.* 2013).

It has been found that *Melipona* sp. starts collection of pollen from watermelon at 0830 h and their foraging activity reached its peak at 1030 h (Bhambura 1958). The colonies of *Trigona itami*, *T. moorei* and *T. minangkabau* with populations of around 5,000, 2,000 and 2,600 made flights of about 7,000, 2400, and 1,200 per day respectively (Inoue *et al.* 1985). In stingless bees, the flight activity depends on species, population of colonies as well as availability of resources. Likewise, workers in a *Tetrgonula carbonaria* hive with 10,000 approximate members completed about 20,000 flights per day (Heard and Hendrikz 1993). A newly established hive of *T. minangkabau* made just 700 flights per day where around 350 workers were present in the hive. We can therefore, summarize that the relationship between hive population and flight activity is positively corelated (Kakutani *et al.* 1993).

Among the honeybees, *A. florea* spent the maximum time (14.63 sec) in collection of pollen followed by *T. iridipennis* (12.89 sec), *A. cerana* (7.59 sec), *A. mellifera* (6.77 sec) and the lowest in *A. dorsata* Fab (Eswarappa 2001). Prakash (2002) observed that the time spent by *A. florea* in collection of nectar from both male and female cucumber flowers was found maximum (305.93 sec. on pistillate and 276.68 sec on staminate flower), followed by *T. iridipennis* (286.61 sec on pistillate and 271.99 sec on staminate flower), *A. mellifera* (37.47 sec on pistillate and 34.00 sec on staminate flower) and *A. cerana* (38.12 sec on pistillate and 35.31 sec on staminate flower). The lowest time spent was recorded in *A. dorsata* (3.52 sec on pistillate and 31.44 sec on staminate flower). He also reported that *A. florea* spent maximum time (13.49 sec) in collection of pollen from cucumber followed by *T. iridipennis* (11.44 sec).

Most of the stingless bee species have a smaller and shorter foraging range than that of the honeybees which may enhance foraging efficiency in confined spaces (Kakutani *et al.* 1993). Free (1993) also reported that the Africanized and Irai stingless bees began to visit flowers at approximately 07:00 h, and visits became more frequent between 10:00 and 12:00h. Observation on foraging activity of two stingless bee species *M. beecheii* and *T. angustula* recorded that 93% of *M. beecheii* foraged from 05:00 to 08:00 h, while 75% of *T. angustula* bees foraged from 08:00-10:00 h and both species were found to collect pollen in the morning. The diurnal and seasonal variations in foraging activity suggested that there was little competition for pollen between the two species (Beismeijer *et al.* 1998).

Brujin and Sommeijer (1997) worked on foraging activity of different stingless bee species namely *Melipona fasciata*, *M. beechii*, *M. favosa* and *T. angustula*. In the studies it was concluded that *Melipona* sp. Followed different foraging pattern and collects pollen early in the day and the foraging activity was found shorter in *T. angustula* species as compared to *Melipona* sp.

Devanesan *et al.* (2002) studied foraging activity of *T. iridipennis* at Kerala and reported that the peak foraging activity was in the month of July and least foraging activity during December and January. Bees started foraging at 07:00 h which gradually increased reaching its peak at 1200 h, but almost no activity at 18:00 h. At 13:00 h there was a decline in foraging activity and increased to second peak at 15:00 h. During first peak at 12:00 h, the average pollen foragers were 17.9 and nectar gatherers were 11.5 while during second peak at 15:00 h, the average pollen and nectar gatherers were 13.9 and 19.8 respectively. The maximum pollen foraging activity was observed in the morning and average nectar foraging activity was observed maximum at mid day.

Worker bees began their usual foraging activities as early as dawn and ended by dusk depending on weather conditions and availability of flora. Peak foraging times happen simultaneously in dry weather when the forage sources are abundant (Kwapong *et al.* 2010). Danareddi *et al.* (2011) reported that the foraging behaviour of *T. iridipennis* varied significantly at different hours of the day and month of the season. Only one peak activity of outgoing and incoming bees with and without pollen occurred between 10:00 and 12:00 h during all the observed seasons. When the foraging activity was taken into consideration irrespective of seasons the peak activity of outgoing bees, pollen foragers and incoming bees without pollen occurred at around 12:00 h (31.40, 10.52 and 17.18 bees/5 min, respectively). The activity of outgoing bees was higher in October and November while that of pollen foragers was notable in February. Similarly in October, November and February to May, a higher activity of incoming bees without pollen was documented during the study.

Hilario and Imperatriz (1999); Hilario *et al.* (2012); Figueiredo *et al.* (2013) also noticed that the foraging activity of stingless bees was lower in winter months and during cold days, the flight activity was less intense and then became negligible during evening hours. This may be due to the increase in stickiness of the resin source once the temperature and sunlight increases, which makes the task more difficult for the bees to collect and load the resin in their corbicula.

Nunes-Silva *et al.* (2013) divulged that the frequency of flower visitation was highest in the morning and concurred with the number of bees exit the hive to forage for pollen. The *M. fasciculata* foragers visited eggplant flowers for 147±16 sec. and the average time spent on a flower (visit duration minus time spent on grooming) was recorded as 96 ±12 sec. During the foraging visits, the bees performed 68±10 buzzes. Inside greenhouses, the flight activity of *M. fasciculata* was busiest in the morning between 06:00 h - 11:00 h which significantly coincided with the opening time of eggplant flowers and anthers from 06:00 to 17:00 hours. This feature between colony foraging pattern and floral biology indicates *M. fasciculata* as promosing pollinator of eggplant flowers.

Nicodemo *et al.* (2013) observed that the floral handling time was long in female flowers compared to male flowers. Bonfim (2014) reported that the stingless bees primarily seek nectar in both pistillate and staminate flowers although during their visits to staminate flowers, they got their bodies dusted with a great amount of pollen grains. A portion of these pollen grains was transferred to the pollen basket and later taken to the hives but a certain amount of pollen grains remained on their bodies, mainly over the head, legs, and thorax and abdomen area.

The floral handling time of female flowers was long in cucumber which produces more nectar compared to male flowers which produce pollen (Tej *et al.* 2017). Stingless bees visited female flowers mostly during 11:00 AM - 12:00 PM and visited male flowers mostly at 11:00 AM (Sawatthum 2017). In muskmelon, stingless bee, *T. laeviceps* do not forage for pollen and nectar during 06:00 h - 08:00 h (Gadhiya and Pastajia 2019). The foraging activity of *Scaptotrigona* sp. Foragers on watermelon flowers usually began at 5:30 am. Although, only 2.46 numbers of bees flew out of their hives at this time, the numbers increased rapidly with a maximum peak at 08:00 h (27.71 bees). The foragers of *Scaptotrigona* sp. secured maximum visit peak during the optimal period of stigma receptivity and pollen viability. Besides moving around the flower, *Scaptotrigona* sp. foragers made multiple visits to the mini watermelon flower. Each flower received an average of 6.0 ± 3.69 visits (n = 24) over a period of five minutes (Bonfim *et al.* 2014). Vijayan *et al.* (2018) observed that both pollen and nectar foraging activity of *Tetragonula iridipennis* was least during the early morning (07:00 h - 09:00 h), whereas resin foragers started their foraging activity as early as 07:00 h - 08:00 h. The pollen foraging activity attained its peak from 8:00 AM to 12:00 PM, resin foraging activity attained its peak from 8:00 AM to 11:00 AM but nectar foraging was observed more towards noon and evening hours (12:00 PM to 5:00 PM).

Stingless bees are efficient pollinators having relatively higher abundance in cucumber, ash gourd, watermelon and brinjal (Chauhan *et al.* 2019; Chauhan *et al.* 2020). The foragers start foraging in late morning hours and the activity ceased in early evening hours.

Among insects, bees are the main pollinators of crops grown throughout the globe for food. Honeybees, stingless bees, bumble bees are important pollinators that depends solely on flowers for their development. Similarly, plants are co-evolved with bees and became inseparable with each other. Many insect visitors forage on the floral resources presented to them by the plant but only a few are good pollinators. In depth information on actual or potential role of individual pollinator species is necessary to compare the potential value of different pollinator species to be candidates for introduction as managed pollinators (Corbet *et al.* 1991). Bees are the most important pollinators and so understanding the scale at which they forage has important ecological implications and conservation applications (Greenleaf *et al.* 2007). Chaudhary and Gupta (2000) have determined the variation in bee activity at hive entrance. The different pollination attributes several characters like relative abundance, foraging rate, foraging speed and loose pollen grains discussed by several researchers (Chauhan 2011, Chauhan and Thakur 2014; Chauhan 2015; Chauhan *et al.* 2019; Chauhan and Singh 2020) revealed the importance in pollination effiency index (Bohart and Nye 1960; Sihag and Rathi 1994). After assigning the pollination index to different pollinators visiting a particular crop, the pollinator having higher pollination index is suitable to be selected as a pollinator for that particular crop (Chauhan 2015).

Honeybees as pollinator of different crops

Honeybees and flowering plants are mutually dependent upon each other for their existence. The majority of plants are reliant on insects for their pollination requirements while insect in turn, depend on plants for provision of energy to nourish and sustain their activities (Seeley 1985). This particular energy relationship between plants and nectar collecting insects is the required foundation for studying the crop pollination, honey production and their foraging strategies (Morse 1976; Akaratanakul 1987).

Many of the agricultural crops like *Aeschynomene americana* L., *Ageratum conyzoides* L., *Amomum xanthioides* Wall, *Anacardium occidentale* L., *Antigonon leptopus* Hook., *Balakara baccata* Roxb., *Castanopsis acuminatissima* Rehd., *Cinnamomum kerrii* Kosten, *Coccinia grandis* C. L.Voigt, *Cocos nucifera* L., *Coffea arabica* L., *Conyza sumatrensis* Retz are pollinated by honeybees (Suwannapong *et al.* 2011). Honeybees are so important as pollinators as they actively seek out flowers with pollen, not only for nectar unlike other pollinators such as bats and humming birds that primarily visit flowers for nectar. Some plants will not produce fruit without honeybee pollination. This floral fidelity of bees is due to their preference for nectars having sugar content and pollen with higher nutritive values.

Asia is bestowed with a rich diversity of honeybee species. These include *Apis cerana*, *A. dorsata*, *A. florea*, *A. laboriosa*, *A. breviligula*, *A. binghami* and *A. andreniformis* that are indigenous to the region, whereas the European honeybee, *A. mellifera* was introduced to the region and promoted for commercial beekeeping. The indigenous honeybees make a significant contribution to the livelihoods of the rural poor and protection of the environment through a variety of products and services (Partap 1992).

The benefit of honeybees as the main group of pollination services for enhancing crop yields and maintaining biodiversity is thought to be much higher than their hive products (McGreger 1976; Crane 1991; Free 1993; Partap and Verma 1994; Suwannapong *et al.* 2011). The process of keeping bees for pollination has shown to increase the crop yield and quality while reducing fruit drop. Honeybee pollination has been reported to increase the fruit juice and sugar content in citrus fruits, reduces the percentage of misshapen fruits in strawberry and increases the oil contents in rape seed and sunflower (Partap 1992).

Qualities of honeybees which make them good pollinators

Honeybees have some important features which make them reliable pollinators of crops. The body is covered with hairs and have structural adaptation for carrying nectar and pollen. They land on the flowers, takes nectar and pollen but do not injure them. The adults and larva feed on nectar and pollen and also store both nectar and pollen for their future use. There is no diapauses and as such food resources are required by them throughout the year. The body size and proboscis length is suitable to collect nectar for many crops. They can forage on different crops with high fidelity in varied weather conditions.

Polylecty, fidelity and consistency are the three major attributes of honeybees which make them better pollinators as compared to other insects.

Under open pollinated conditions, with honeybee pollination the yield and quality of the crop increases manifold both in self-pollinated and cross-pollinated crops (Table 13.2 and 13.3). Although, honeybees are the major pollinators throughout the world but due to the increase in the diseases, pests, enemies and climatic conditions, they are not alone able to complete the present requirement of pollinators in the crops (Delplane 1995). So the researchers are searching for the development of rearing technologies for other pollinators like bumble bees and stingless bees.

Moreover, the agriculture is getting intensive and the burden to grow more crops from small land is increasing. Farmers are gradually shifting towards greenhouse cultivation. In greenhouses, honeybees are not able to forage and colony losses are reported. But other pollinators such as bumble bees, stingless bees and xylocopa are able to pollinate the crops because of certain required characters present in them. Bumble bees are emerging as a promising pollinator for pollination of crops grown under protected conditions especially those that require buzz pollination (tomatoes). In temperate conditions and in automated environment-controlled greenhouses, bumble bees are showing excellent results as a pollinator and increasing the quality and production of crops.

Bumble bees as pollinator of different crops

Bumble bees are also highly developed social bees like honeybees though they possess smaller, less structured nests that may consist of up to five hundred bees in a colony. These are the most diverse group among all the

Table 13.2. Crops benefitted by bee pollination

Fruits and nuts	Vegetable and vegetable seed crops	Oil seed crops	Forage seed crops
Almond	Cabbage	Sunflower	Lucerne
Apple	Cauliflower	Niger	Clover
Apricot	Carrot	Rape seed	
Peach	Coriander	Mustard	
Strawberry	Cucumber, Melon	Safflower	
Citrus	Onion, Pumpkin	Gingelly	
Litchi	Radish, Turnip		

Table 13.3. Per cent increase in yield due to bee pollination

Crop	Yield (%)
Mustard	43
Sunflower	32-48
Cotton	17-19
Lucerne	112
Onion	93
Apple	44

pollinators in temperate regions and it may be noticed that only 34 of the 250 species occur in the tropics (Williams 1998). In comparison to honeybees, bumble bees work harder, longer and faster even at cooler temperatures (Light 1994). These bees are also known as buzz bees which produce high frequency vibrations by beating of wings and are considered to be very efficient pollinators of cucurbits and solanaceous crops under open as well as protected cultivation (Paydas *et al.* 2000a; Thakur 2006). The bumble bees are commercially employed in the pollination of crops grown under protected conditions. These bees pollinate many crops like tomatoes, egg plants, peppers, melons, raspberries, blackberries, strawberries, blueberries and cranberries (Smith 1998). These are the only pollinators of potato flowers worldwide and the main pollinators of many high value crops (Parkar and Torchio 2007). Generally, they are more efficient and reliable pollinators especially under protected conditions over honeybees as they carry a considerable amount of loose pollen grains, have short flight range and less population (Mackenzie 2009). They usually establish themselves in underground nests which are made up of different nesting materials such as grass, leaves, wood pieces with one or more entrances (Hines *et al.* 2007). These can be reared artificially (Goodell and Thomson 1997).

Bumble bees were proven to effectively to pollinate various crops like red clover, cucumber, watermelon, cotton, tomato, pepper and strawberry effectively. Hobbs *et al.* (1961) and Holm (1966) established that the bumble bees are commendable pollinator species for pollination of flowers with deep corolla. Bumble bees have been acknowledged as one of the most important pollen vectors in cotton (Trelease 1879; Thies 1953; Weaver 1978). The study was conducted to evaluate the role and potential of the bumble bees in high hybrid seed production in sterile cotton in Texas and also found that bumble bees carries more pollen than *Apis* sp. foragers and therefore, could provide a potential pollination alternative in sterile hybrid cotton. Honeybees can also pollinate tomato, pepper, egg plant, melon, watermelon, cucumber, strawberry, crane berry etc, but are less efficient than bumble bees (Free and Butler 1959; Holm 1966).

Bumble bees are beneficial pollinators especially in pollinating those flowers whose sizes facilitate pollen transfer while visiting the nectaries (*Gossypium* sp. and certain cultivars of tree fruits) in pollinating flowers with narrow deep corolla tubes from which only insect with long tongue can obtain nectar (*Vicia faba*). In 'buzz pollinating' crops (some *Actinidiaceae*, *Boraginaceae*, *Aricaceae* and *Solanaceae*) where the high frequency wing vibration causes the release of pollen often in large amounts to settle onto the surface of the bee's body under the influence of an electrostatic field that originates from its body (Corbet *et al.* 1982, 1988; Ericson and Buchmann 1983).

Macfarlane *et al.* (1983) reported that the red clover seed yield was 210 kg/ha when bumble bee queens were released and it decreased to 100 kg/ha when these queens were not available in the next two years. Bumble bee pollination yields higher fruit quality and increase in total yield (Banda and

Paxton 1991; Abak *et al.* 1995). Bumble bees could solve the purpose of a backup or alternate pollinator to honeybees for pollinating cucumber, watermelon and possibly other vine crops grown in either open or greenhouse conditions (Stanghellini *et al.* 1997).

Pomeroy and Fisher (2002) noticed that bumble bee pollination can increase the fruit quality in Kiwi by visiting in large numbers which in turn, transfer more amount of pollen to the female flowers and hence the quality of fruits is improved. Hence, the fruit growers benefit from bumble bee pollination because of lower production cost, increased yields and improved fruit quality in many vegetable, fruits and seed crops (Velthius and Doorn 2006). Bumble bees are among the most efficient pollinators not only for the wild plants but also for pollination services, used in both outdoor and greenhouse horticulture and orchards (Wolf and Moritz 2008). Use of bumble bee as a pollinator in pollinating crops was found cheaper (9,100 pounds) than mechanical pollination (10,000 pounds) in Netherlands and Belgium (Roulston and Julier 2009).

Bumble bee pollination in protected cultivation

The need for pollination of crops in enclosures arises to produce noncontaminated seeds or to find whether the production and productivity of crop increases. Bumble bees work well in confinement and are especially valuable for use in small enclosures. Bumble bees used for pollinating *Trifolium prantese* confirmed that these were effective in seed production even in glass houses. Neiswinder (1956) found that honeybees are not effective pollinators of solanaceous crops and bumble bees used for the pollination of tomato in greenhouse increases the yield and productivity. In Newzealand, the production of red clover seed was dramatically increased with the introduction of four bumble bee species (Holm 1966). Bumble bees can be used to pollinate various solanaceous and fruit crops in greenhouse (Morgner 1977).

The pollination of muskmelons (*Cucumis melo*) in greenhouse using *B. terrestris* was evaluated. Nine hives of bumble bees were kept in 1500 m^2 green house and the bees were allowed to forage on melon flowers. The male flowers were larger in number than female flowers. It was found that 90% of the crop attained the required exportable weight and thus, it was concluded that bumble bees are an effective mode of pollinating greenhouse melons (Richard and Pomeroy 1989).

Four species of bumble bees were used *B. pascorum, B. terrestris, B. pratorum* and *B. hyponorum* for the pollination of primula (seed crop) in the green house. In Primula hand pollination was recommended but they found that *B. pascorum* pollinated the crop efficiently and the production was 50% more than hand pollination (Velthius and Laura 1991). In Netherlands, artificial vibrators were used for the pollination of tomato grown under greenhouse or glasshouse. However, once bumble bees were introduced in the glasshouse for the pollination of tomato crop and an increase in the crop yield and productivity of crop was recorded, whereas in the control condition the crop yield, size and weight was much lower than bumble bee pollinated crop (Ravestigin and Sande 1991 ; Asada and Ono 1996).

Richards and Myers (1997) used the bumble bee *B. accidentalis* for the pollination of *Cicer milkvetch* and concluded that the crop closer to the nest had more percent seed set than the control plots where no bumble bees were used. Bumble bees and honeybees were evaluated for the pollination of crops grown in polyhouses and it was found that both bumble bees and honeybees increased the yield in greenhouse but the yield, fruit weight, diameter and number of seeds were higher in bumble bee pollinated crops than honeybee pollinated crops (Abak *et al.* 1999).

Bumble bees are more efficient pollinators of strawberry grown under plastic greenhouses in cold cloudy weather conditions (Paydas *et al.* 2000). In Blueberry, pollination through bumble bees increased percent fruit set, percent fruit weight and number of seeds/berry (Constance and Francis 2001). In *Vaccinium* crops, honeybees are not the reliable pollinators as these crops need vibration to shed pollen and bumble bees are able to perform vibration actions during foraging and as such, they can be exploited commercially to pollinate the *Vaccinium* crops under greenhouses (Mackenzie 2009).

Attal *et al.* (2003) studied the effect of bumble bees, plant growth bio-regulators, hand vibration and control on tomato production in greenhouse. They discovered that bumble bees produced a yield which was significantly higher than plants treated with plant growth bio-regulators, hand vibration and control. *B. impatiens* is a better pollinator for crops grown under protected conditions (Whittington and Winston 2004). Bumble bees were used to pollinate two varieties of sweet pepper (Bardenas and Vargase) in greenhouse and the average percentage of seed set per fruit of the two varieties were 49.8% (Bardenas) and 40.7% (Vergase), whereas the averages of the control treatment were 27.5% and 25.7%, respectively (Sanz and Serrano 2006).

One colony of bumble bee, *B. haemorrhoidalis* was kept in the plastic greenhouse for the pollination of cucumber crop and was found that the bumble bee foragers visited 59.5, 51.0 and 45.5 flowers per five minutes during 0600 h, 0800 h and 1600 h (Thakur and Kashyap 2008). Chauhan (2011) found bumble bees activity started early in the morning and ceased in late evening hours. During the noon hours, bees usually do not forage but remain inside the hive.

Effectiveness of bumble bees as pollinator

Most of the crops are dependent on insect pollinators (McGregor 1976). Bumble bee is regarded as one of the most efficient pollinator of many crops and can forage even at low temperature where honeybee activity is limited and are valuable in pollinating those large sized flowers which facilitates pollen transfer during their visit to the nectaries (*Gossypium* spp.), in pollinating flowers with deep corolla tubes (*Trifolium prantese*), and in buzz pollinating crops (Actinidiacae, Solanacae) where the high frequency wing vibration causes pollen to be released, often in large amounts (Corbet *et al.* 1982; Erikson and Buchmann 1983).

Bumble bees are excellent pollinators of red clover and many other seed crops in Scandinavia where weather conditions are hostile to honeybees (Bohart 1972). It was recorded that more number of bumble bees were observed visiting field beans and pollinating more number of flowers during a single visit as compared to honeybees (Poulson 1973).

In pollination under greenhouse conditions, bumble bees are rightly suggested as an alternate pollinator to honeybees as they do not disorient from the main crop even when windows are open and can be used to pollinate the crops under low temperature where honeybees are not able to pollinate the crop and also, they do not look for alternate sources outside the greenhouse (Heemert *et al.* 1986). On *Cirsium altissimum,* i.e. meadow rose, bumble bees have been observed taking more pollen from the flowers than honeybees (Plowright and Laverty 1984). Free and Ferguson (1980) reported that the bumble bee *B. teresitris* and *B. pratorum* were more abundant than other pollinators (honeybees and syrphids) on *Brassica napus* L. (oil seed rape) and the seed production was increased considerably.

Pear requires cross pollination and is self unfruitful. It is pollinated by bumble bees and honeybees. The honeybees are more in number than bumble bees found on the crops. This happens as most of the bumble bee queens started rearing brood during that period therefore, population of bumble bees is less (Mayer *et al.* 1987). The bumble bees are better adapted to pollinate the greenhouse crops (melons, tomato) under adverse conditions and are less inclined to look for the alternate sources of pollen and nectar outside the greenhouse.

Berger *et al.* (1988) noticed that bumble bee (*B. pennsylvanicus*) is a potential pollinator of cotton in Texas and can be used for hybrid seed production. The foraging behavior and pollination efficiency of *B. pennysylavanicus* and *A. mellifera* were compared and was observed that *B. pennysylavanicus* collected pollen and nectar with a large number of pollen grains found on the body of bumble bees than *A. mellifera*. They concluded that bumble bees are a more potential alternative to traditional methods of pollinating MS (male sterile cotton). In Mediterranean coastal regions malformation of fruit produce in strawberries was prevented by bee pollination method (Ahn *et al.* 1988). Growers should introduce bumble bees to their greenhouses especially in winter months to get early, high yield, large and regular shaped strawberries (Pinzauti 1993). Rhododendrons also require vibrations for the release of pollen from the flowers and so bumble bees were made to visits these flowers in large number. It was later found that some amount of pollen was regularly attached in the corbicular legs (King and Buchmann 1995).

Bumble bees and honeybees are very useful in early production of strawberries but bumble bees were very efficient in pollination at cold weather (Paydas *et al.* 2000). In field studies, pollination effectiveness of the commercially reared bumble bee *B. impatiens* Cresson was compared to the honeybee *Apis mellifera* for low bush blueberry, *Vaccinium augustifolium*. It was observed that per cent fruit set, per cent harvested berries, berry weight and seeds per berry were higher in case of *B. impatiens* and was found to be a suitable alternative to honeybees as pollinator (Stubbs and Drummond 2001).

Bumble bees are a viable alternative to honeybees or hand pollination in the production of greenhouse grown musk melons and they can continue to forage in overcast weather unlike honeybees (Fisher and Pomeroy 1989). For the pollination of round tomatoes, bumble bee colonies were introduced in the glasshouse. In the absence of bumble bees, vibration is necessary to obtain good fruit set and yield, but when bumble bees are introduced in crop, vibration makes no difference and the fruit set of 96% was observed and the vibration was suggested to be completely removed (Ravestijin and Sande 1991). Bumble bees act as an excellent pollinator of high value crops under glass house and are not troubled with parasite problems compared to the devastation caused by *Varroa* mite in honeybees and thus can be a backup pollinator for honeybees (Delplane 1995). Several studies on effectiveness of honeybee and bumble bee pollination on fruit set and abortion of cucumber and watermelon have revealed that cucumber flowers visited by *B. impatiens* had lower percent fruit abortion than *A. mellifera* when they were compared with equal bee numbers (Stanghellini *et al.* 1997). It was noticed that the yield remained the same in both honeybee and bumble bee pollination, but the fruit weight, height, diameter and number of seeds were higher in bumble bee pollinated fruits than honeybee pollinated ones (Dasgan *et al.* 1999).

Dasgan and Ozdogan (2004) and Attal *et al.* (2003) concluded that in tomato greenhouses, bumble bee pollination should be used for increased yield and more marketable fruits instead of growth regulators and vibration as the results revealed that bumble bee pollination increased the yield by 90% and 61% over growth regulators (41%) and vibration (9%). In greenhouse hot pepper *Capsicum annuum*, it was found that with the introduction of bumble bees, the fruit mass and number of seeds increased by 27.2% and 47.8% (Kwon and Saeed 2003).

Higo *et al.* (2004) studied the benefit of *A. mellifera* honeybees as a supplement to bumble bee and observed that there was a remarkable increase in the fruit size of tomato that were grown inside greenhouses provided with bees. In another study, the bumble bee activity was documented on sweet pepper to ascertain the fruit quality in greenhouses and discovered that there was a significant increase in the seed set per fruit to the time or 49.8% or 40.7% more than those under control (27.5% and 25.71%) in two varieties (Sanz and Serrano 2006).

Tung and Kong (1997) meticulously studied the pollination efficiency of bumble bee (*B. terrestris*) in greenhouse. They introduced two bumble bee colonies in the tomato greenhouse and observed higher fruit set, number of tomatoes produced, size, sugar content and weight. It was concluded that utilization of two bumble bee colonies in 50 × 50 m greenhouse for tomato pollination during the cool season could totally substitute other more laborious pollination methods. All *Vaccinium* crops do not require cross pollination but for good fruit quality, vibration is required. Bumble bees employed for this purpose helps in increasing productivity due to buzz pollinating the crop and is more effective than other pollination methods (Mackenzie 2009).

In India, different researchers worked on bumble bee rearing (Dayal and Rana 2004; Thakur *et al.* 2005; Kashyap 2008; Chauhan 2011; Chauhan 2015) and its utilization in crop pollination (Chauhan and Thakur 2014; Nayak 2018; Thakur 2018; Yankit *et al.* 2018). Chauhan and Thakur (2014) conducted an experiment to artificially rear the native bumble bee and assess the pollination potential of bumble bee, *Bombus haemorrhoidalis* Smith in Himachal Pradesh, India (Fig. 13.2). He found bumble bees increased the quality and quantity of the cucumber under greenhouse conditions (Table 13.4).

Thakur (2018) reported an increase in bell pepper pollinated by bumble bee, *Bombus haemorrhoidalis* Smith under protected conditions (Table 13.4) in Himachal Pradesh (Fig. 13.4).

Nayak (2018) reported an increase in the production and quality of kiwi when pollinated by bumble bees as compared to other pollinators and hand pollination (Figs. 13.3 and 13.5).

Fig 13.2. Bumble bee colony for pollination in cucumber under greenhouse and bumble bee pollinating cucumber (*Courtesy:* Avinash Chauhan).

Table 13.4. Impact of bumble bee pollination on cucumber and bell pepper grown under protected conditions

Quality parameter	Cucumber (%) increase	Bell pepper (%) increase
Fruit set	27.64%	25.00
Healthy fruits	22.81%	23.84
Crooked fruits*	65.62%	38.30*
Fruit length	20.78%	13.51
Fruit diameter	23.80%	21.52
Fruit weight	36.48%	24.60
Seed number	21.52%	113.65
Weight (1,000 seeds)	12.54%	5.44

*Per cent decrease

Fig 13.3. Bumble bee colonies for pollination of kiwi and worker collecting pollen on kiwi (*Courtesy:* Rohit Nayak).

Fig 13.4. Bumble bee colony for pollination in bell pepper under greenhouse and bumble bee pollinating blooms (*Courtesy:* Priyanka Thakur).

The effectiveness of the bumble bee *Bombus haemorrhoidalis* Smith (Hymenoptera: Apidae) was evaluated and compared to non-bumble bee pollinated tomato (*Solanum lycopersicum*) crop grown under protected conditions. Likewise, bumble bee colony reared in laboratory and utilized for pollination of tomato cultivar (Solan Lalima) grown inside polyhouse resulted in increased number of fruits per cluster (6.76 fruits/cluster), number of fruits (75.80 fruits/plant), fruit length (5.16 cm), fruit breadth (5.75 cm), fruit weight (93.87 g), fruit yield (12.7 kg/m^2) and healthy fruits (90.33%). There was also reduction in number of misshapen fruits (9.8%), number of seeds (102.95 seeds/fruit), 1,000 seed weight (6.32 g) over control crop in comparison to bumble bee pollinated crop. Bumble bee pollination accounted per cent increase in number of fruits per plant, healthy fruits, fruit length, fruit breadth, fruit weight, fruit yield, number of seeds and 1,000 seed weight by 38.41, 21.94, 46.45, 50.82, 57.66, 64.79, 78.54 and 78.80% , respectively (Yankit *et al.* 2018, Fig 13.6).

Four species of bumble bees were reported in Himachal Pradesh. These were *Bombus haemorrhoidalis, B. assmensis, B. rufofasciatus* and *B. trifasciatus*.

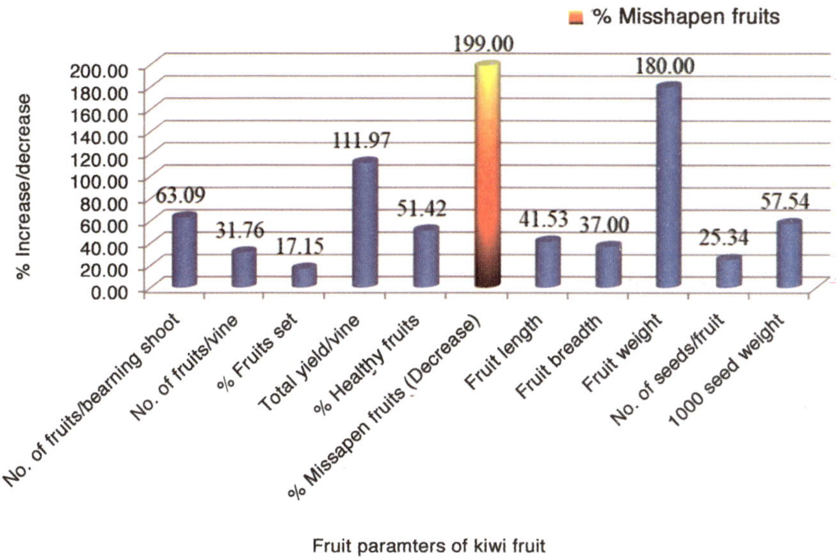

Fig 13.5. Impact of bumble bee, *Bombus haemorrhoidalis* pollination on kiwi (data shown by red bar is reduction in deformed fruits).

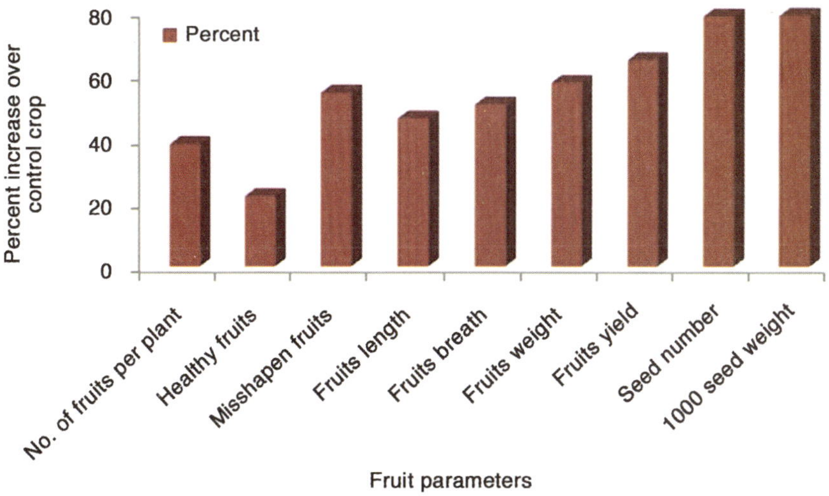

Fig 13.6. Impact of bumble bee, *Bombus haemorrhoidalis* pollination on tomato as per Panma *et al.* (2018).

Out of all these species, *B. haemorrhoidalis* is the only one which is explored for pollination potential and artificially reared under laboratory conditions round the year. The year round rearing by Chauhan *et al.* (2014) and efforts made in bumble bee rearing including morphometrics, molecular identification, caste detemination, biology, karyology, field establishment, pests and diseases and pollination (Thakur *et al.* 2010; Chauhan *et al.* 2013; Chauhan *et al.* 2013; Chauhan *et al.* 2013; Chauhan *et al.* 2015; Chauhan *et al.* 2015; Chauhan *et al.* 2016; Chauhan *et al.* 2017) laid the foundation for the commercialization of bumble bees in India.

Constraints in bumble bee pollination

- Bumble bees are not able to forage at higher temperatures. In India, majority of greenhouses are naturally ventilated. The temperature inside the greenhouses reaches high which restricts the bee activity and thus, the required pollination cannot be attained.
- Bumble bee rearing is still under research and the technology is not commercialized so fetching bumble bee colonies for pollination is still a big challenge.
- If imported, there might be chances of pest insurgency and negative impact on native bumble bee species.

***Xylocopa* as pollinators of crops**

Xylocopa is commonly found in less disturbed areas and are also known as carpenter bees because of their characteristic nesting behaviour. They are distributed mainly in the subtropical and tropical regions of the globe but a few species are also documented from the European countries. The large carpenter bees belong to the tribe Xylocopini within the subfamily Xylocopinae (Hymenoptera: Apidae). Currently, these bees are grouped into a single genus, *Xylocopa*. More than 500 species are available worldwide and out of which, 40 of them are reported from India. *Xylocopa* exhibits both solitary and social behaviour. The size of the adults is similar to bumble bees but they do not possess bands on their abdomen. Some species have coloured pubescence on dorsal side of thorax. These are active between February to October and in winter the adult queens/males undergo hibernation. The bees are generally dark metallic in colour and polymorphism is exhibited by different species. The social *Xylocopa* species have queen, males and few workers which can be distinguished.

The large carpenter bees (genus *Xylocopa*) are wood-nesting generalist pollinators under a broad geographical distribution that exhibit varying levels of sociality. Their foraging activity is characterized by visits upon a wide range of food plants, long season of activity, tolerance of high temperatures and activity even under low illumination levels. These natural traits drive them as fascinating candidates for agricultural pollination in hot climatic areas, particularly in greenhouses including night-blooming crops. Carpenter bees have exhibited competent pollination service in many crops like passion flower, blueberries, greenhouse tomatoes and greenhouse melons (Keasar 2010).

Pollination potential

Carpenter bees pollinate passion flower (*Passiflora* spp.) in their native habitats (Mcguire 1999) as well as in commercial agricultural settings (Corbet and Willmer 1980; Siqueira *et al.* 2009). They provide better pollination service than honeybees for this crop (Roubik 2018). *Xylocopa* subgenus Lestis has been successfully reared in greenhouses for tomato pollination in Australia. Their foraging activity led to an increase in tomato weight by 10% relative to a combination of wind and insect pollination. The efficiency of carpenter bees in pollinating tomatoes is increased by their ability to buzz the anthers (Hogendoorn *et al.* 2000). In a pilot study in Israel, it was revealed that the fruit set of greenhouse-grown honeydew melons was three times higher when pollinated by *X. pubescens* compared to honeybee pollination (Sadeh *et al.* 2007). Social and solitary nesters had similar efficiency in pollinating this crop as they did not differ in the daily activity patterns and flower visit rates. Pollination by both types of nesters led to similar fruit sets, fruit mass, and fruit seed number (Keasar *et al.* 2007). These carpenter bees are important pollinators of cotton in Pakistan, India, and Egypt (Watmouth 1974). *X. varipuncta* is compared favourably with honeybees (*Apis mellifera*) as pollinators of male-sterile cotton in field cages in the USA (Waller *et al.* 1985).

In India, research trials on pollination are going on in Bihar, Assam and Nagaland to establish the pollination efficiency of this pollinator. Different species like *Xylocopa fenestrata*, *X. leucothorax* and *X. tenuiscapa* are under experiments on pollination efficiency for exploration as pollinators under protected conditions (Fig 13.7).

Prospects of *Xylocopa* as a pollinator

Keaser (2010) explained the qualities of *Xylocopa* genus for consideration as a pollinator. Among different bee species, many bees have a short activity season and are specialist foragers and therefore do not provide a broad alternative to honeybee pollination. On the other hand, carpenter bees have long activity seasons and feed on a wide range of plant species. In addition, they are capable of buzz pollination like bumble bees. This makes them potentially more versatile as agricultural pollinators. Hibernation occurs in the adult stage and females start foraging whenever temperatures reach high enough values. This means that it is relatively easy to manipulate the onset of foraging in greenhouses. Another important advantage is that the genus has a worldwide distribution. This implies that local species of *Xylocopa* can potentially be used over wide areas, reducing the need to import exotic pollinators. The possibility to lure these bees into suitable artificial nesting material allows provisioning of nesting material that can be easily used in agricultural settings and moved to places where pollination services are needed (Hogendoorn 2004).

Constraints in Xylocopa bee pollination

The current challenges in the commercialization of these attempts lie

Xylocopa bee pollinating cucumber

Xylocopa frame

Fig 13.7. *Xylocopa* spp. pollination in cucumber and nesting frame designed for pollination (pc. Ataur Rehman and Mukul Dieka, India)

in the difficulties of mass-rearing *Xylocopa*, and in the high levels of nectar robbing exhibited by the bees (Keasar 2010). Scientific hives for *Xylocopa* rearing are still not developed in the country and as such, it requires lots of research before commercialization.

Stingless bees as pollinators of crops

Most of the crop plants depend on pollination for fruit and seed set. Stingless bees (Apidae, Meliponini) have received increasing attention as crop pollinators in Latin America, during the past few years. These highly developed eusocial bees live in perennial colonies and are easily domesticated in hives. They possess various behavioural traits which are distinct, such as recruitment of foragers (Lindauer and Kerr 1960), high flower constancy, great diet-breadth, and easy adaptation to new plant species that favours them as auspicious contenders in pollination of commercial crops (Roubik *et al.* 1986; Ramalho *et al.* 1994; Nogueira-Neto 1997; Heard 1999).

Stingless bees are known for their expensive honey and pollination potential. Stingless bees have certain characters which make them a true generalistic pollinator. The important traits are defined by various workers (Heard 1999; Roubik *et al.* 1986; Ramalho *et al.* 1994; Chauhan and Singh 2020) which are as follows:

Strengths and limitations of stingless bees for crop pollination as per Heard (1999)

In the words of Heard (1999), many attributes of stingless bees are similar to those of honeybees. Some of the characteristics that impact their capability as pollinators are

(a) **Polylecty and adaptability**: It enable them to pollinate multiple plant species and also adapt to new plant species.

(b) **Floral constancy**: It is a phenomenon where a worker usually visits only one plant species during a single foraging trip (Ramalho *et al.* 1985).

(c) **Domestication**: It is the ability to hive colonies, inspect, propagate, feed, requeen, control enemies, transport and managed methodically (Nogueira 1997).

(d) **Perennial colonies**: It allow workers to forage continuously within the given climatic constraints and obviate the need to develop their respective colonies each year.

(e) **Large food reserves stored in nests**: This trait obviously benefits the colonies to survive long periods of low food availability or dearth. In other words, it means that the workers collect floral resources that exceed immediate needs and it ultimately result in intensive visitation of preferred flowers (Roubik *et al.* 1996).

(f) **Possibility of in-hive pollen transfer**: It decreases the need for movement of bees between plants of self-incompatible species: This character has been found true for honeybees (DeGrandi-Hoffman 1986) and is most likely influential even for stingless bees.

(g) **Forager recruitment:** The process where workers recruit nest mates to rich floral resources and furnish information on the location of cited floral resources, which in turn allow the rapid deployment of large number of foragers (Neih and Roubik 1995) as compared to other bees and insects where each individual has to find the resource.

Advantages of stingless bees as pollinators (Heard 1999)

Stingless bees are generally less harmful to humans and domesticated animals and are able to forage effectively even in glasshouses (Kakutani *et al.* 1993). The propagation of colonies contributes to preservation of biodiversity by conservation of species and their populations that may otherwise decline due to human disruption of natural ecosystems. The colonies rarely abscond as the old queen is flightless (Inoue *et al.* 1984) and they are resistant to diseases and parasites of honeybees (Delfinado-Baker *et al.* 1989). Thus, a honeybee epizootic that may disrupt pollination would not affect the stingless bees in that system.

Disadvantages of stingless bees for crop pollination

There is a poor level of domestication technology for most species of stingless bees and lack of availability of large number of hives. The colony growth rates are low as compared to honeybees and some species of stingless bees are not domesticated because of specific nesting requirements. Some species are known to damage leaves of crops in search of resin (Camacho 1966; Wille 1965) and some species are territorial and fight when placed in close proximity.

Aspects of the biology of stingless bees in relation to pollination

The biology of the stingless bees has been reviewed (Michener 1974; Roubik 1989; Sakagami 1982; Wille 1983) but never from the perspective of crop pollination. Although stingless bees are generalist flower visitors but all studied species visit a broad range of plant species. It has been revealed that the number of plant species visited for nectar may be higher than the number visited for pollen (Ramalho *et al.* 1985). Despite their generalized flower selection, all examined species show preferences (Lobreau-Callen 1990; Ramalho *et al.* 1985; Roubik and Moreno 1990).

Stingless bees are adaptable social insects that can promptly learn to exploit the resources offered by introduced plants. For example, neotropical stingless bees were seen to forage heavily on the introduced *Eucalyptus* spp. (Guibu *et al.* 1988, Kleinert-Giovannini and Imperatriz- Fonseca, 1987). Some generalizations can be proposed in regard to the plant or flower type which are preferred by stingless bees but it has been suggested that they prefer small flowers (Wille *et al.* 1983), dense inflorescences (Roubik and Moreno 1990), flowers with corolla tubes shorter than the bee's tongues (Heard *et al.* 1990), flowers with long corolla tubes that are wide enough for the bees to enter (Engel and Dingemens 1980), trees (Kleinert-Giovannini and Imperatriz-Fonseca 1987; Ramalho *et al.* 1985, Ramalho *et al.* 1989), and white or yellow flowers (Cortopassi-Laurino *et al.* 1991). Floral constancy is a phenomenon in which a worker visits only one plant species on a single trip and it is typical character of many bees (Eickwort and Ginsberg 1980). In a study conducted from Brazil, 97% of the pollen foragers from nine stingless bee species visited only one floral resource on each trip which is corroborated by the pure pollen loads in their corbiculae during the recorded observations (Ramalho *et al.* 1994). The floral constancy of these bees are associated with pollinator effectiveness because collection and deposition of pollen from two or more plant species reduces the amount of pollen available and as such contaminate the stigmas with the wrong pollen. In addition to floral species constancy, foragers often show resource constancy on nectar, pollen or resin within a single trip and usually between successive trips (Sommiejer *et al.* 1983).

Pollination potential of stingless bees in crops

The pollination potential of stingless bees is described by various workers but is compiled thoroughly by Heard (1999). Woo-kunsuk *et al.* (1996) studied foraging activity and brood rearing of *Trigona* sp. in greenhouse and suggested stingless bees are potential pollinators of greenhouse crops. If sufficient food is available even at a high humidity and temperature (>30°C), the bees can be reared successfully.

Heard (1999) reported that some stingless bee species from the genera *Trigona* are effective pollinators of *Cocos nucifera* (Coconut), *Mangifera indica* (Mango), *Sechium edule* (Chayote), *Averrhoa carambola* (Carambola) and *Macadamia intergrifolia* (Macadamia). The vanilla plants *Vanilla planifolia* is hand pollinated and in Mexico one stingless bee species "abeja de monte" of the genera *Melipona*, is reported to be the only natural insect primarily capable of pollinating blossoms of this crop (Herbst 2001).

In Indonesia, Klein *et al.* (2003) reported that four *Trigona* species, contribute a lot to the pollination of coffee (*Coffea arabica* and *C. canephora*) and *Trigona* (*Lepidotrigona*) *terminate* was found to be the most efficient pollinator. Malagodi-Braga and Kleinert (2004) have reported that

Tetragonisca angustula is an effective pollinator of strawberry in greenhouses. In Japan, pollination tests of stingless bees (*Trigona carbonaria* and *Scaptotorigona bipunctata*) and honeybees (*Apis mellifera*) on white clover, tomatoes, cucumbers, eggplants, paprika and red peppers in greenhouse and apiary areas showed that stingless bees pollinate just as honeybees (Kazuhiro 2004).

In Mexico (Yucatan), González-Acereto *et al.* (2006) have reported that pollination efficiency for tomato (*Solanum lycopersicum*) and habanero pepper (*Capsicum chinense*) in the greenhouses by *Nannotrigona perilampoides* and *Meliponula beecheii*, showed that this species of stingless bee could be used as an alternative pollinator to *A. mellifera* or bumble bees (*Bombus* spp.). Evaluation of *Meliponula beecheii* for pollination of habanero pepper in enclosures showed that it was efficient at pollinating this crop.

Slaa *et al.* (2006) reported that some stingless bee species in the genera *Trigona* and *Plebeia* are efficient pollinators of avocado crop. Ish-Am *et al.* (1999) reported that eight to ten species of stingless bees were effective pollinators of avocado, together with the Mexican honey wasp. Can-Alonso *et al.* (2005) detected that *A. mellifera* and *Trigona nigra* carried comparable amounts of avocado pollen grains on their bodies.

Santos *et al.* (2009) studied the pollination efficiency of the stingless bee, *Melipona quadrifasciata* and the honeybee, *Apis mellifera* in tomato plots in Brazil. They revealed that the largest number of fruits, the heaviest and largest tomatoes, and the ones with the most number of seeds were collected from the greenhouses provided with stingless bees. The stingless bee *M. quadrifasciata*, was found to be more efficient remarkably than honeybees in pollinating greenhouse tomatoes. In Mexico, the stingless bee *Nannotrigona perilampoides* species has been reported to be a beneficial alternative pollinator to *Bombus impatiens* in the pollination of habanero pepper (*Capsicum chinense* Jacquin) because no significant differences in the quality of fruit produced was recorded in the greenhouse under tropical climates (Palma *et al.* 2008). In Australia, pollination of green pepper by *Trigona carbonaria* increased fruit weight by 11% and the number of seeds/fruit by 34% in comparison to the crops that were self-pollinated (Occhiuzzi 2000).

In Brazil, Cruz *et al.* (2005) suggested that *Melipona subnitida* can be considered an efficient pollinator of sweet pepper and seed growers could use *M. subnitida* to increase seed production of sweet pepper variety.

On the contrary, Roselino *et al.* (2009) reported that when comparing two stingless bee species, *Nannotrigona testaceicornis* and *Scaptotrigona depillis*, as strawberry pollinators, it was found that *N. testaceicornis* was a superior pollinator to *S. depillis*. This might be due to the foraging habit of the bees for pollen and nectar in which *N. testaceicornis* take longer time to forage than the *S. depilli*.

Namwong (2010) found that high fruit yield of dragon fruit was obtained with stingless bees than open and hand pollination. In addition, Kakutani *et al.* (1993) compared the performance of stingless bee *Trigona minangkabae* with *Apis mellifera* to pollinate strawberries in greenhouse and the results showed that both the bees have similar efficiency in pollinating the strawberries.

In Brazil, *Melipona fasciculata* was evaluated to know its efficiency as pollinator of eggplant (*Solanum melongena* L., Solanaceae) in greenhouses. They discovered that a single flower-visit by the bees was sufficient for precise transfer of pollen to the stigma thereby promoting pollination. Pollination by *M. fasciculata* increased both the yield and quality (weight) of the harvested fruits and thereby showing *M. fasciculata* as an efficient pollinator of eggplants which also suggested that they may be a feasible alternative to bumble bees in Brazil (Nunes-Silva *et al.* 2013)

In Australia, pollination of green pepper by *Trigona carbonaria* increased fruit weight by 11% and the number of seeds per fruit by 34% compared to the crops that were self-pollinated (Occhiuzzi 2000).

As reported by Sarto *et al.* (2005) greenhouse tomatoes which were pollinated by *M. quadrifasciata* were found to produce greater fruit set and higher fruit quality than those without pollination. Santos *et al.* (2009) also revealed that pollination by *M. fasciculata* increased the fruit set of eggplants by 29.5, 32.5, 45.7, and 12.1 % in comparison to the control group without pollination, autogamy, geitonogamy and cross pollination. It was proven that bee pollination notably increased fruit weight by 96 % when compared to the control group and also by 34 % as compared to the autogamy treatment which was evaluated by Nunes-Silva *et al.* (2013).

Stingless bee, *Tetragonula iridipennis* pollinated cucumber plots in TNAU orchard had an average fruit length of 19.1 cm compared to 17.6 cm in control while the pollinated plots in Srivilliputhur had an average fruit length of 18.8 cm compared to 17.3 cm in control. Prominent increase was also observed in the fruit girth, and yield attribute in the pollinated plots. Azmi *et al.* (2017) observed that the cucumbers which were pollinated by stingless bees produced significantly heavier (0.43 kg ± 0.08) and longer fruits (22.20 cm ± 1.12) than the weight and length of cucumbers produced from pollination treatment without stingless bees. Cauich *et al.* (2004) evaluated the pollination efficiency of *N. perilampoides* on greenhouse tomatoes in Subtropical Mexico. He reported that the species is not a buzz-pollinator yet it effectively pollinated tomato plants grown in netted cages (4 × 4 × 3.5 m, one colony for 40 plants). Santos *et al.* (2004) compared pollination effectiveness *of Melipona quadrifasciata* and *Apis mellifera* (each species in a 86 m^2 greenhouse), and found that tomatoes were bigger, heavier and had more seeds following pollination by *M. quadrifasciata* when compared to *A. mellifera*. The native bee, *Nannotrigona perilampoides* and bumble bee has also been evaluated with promising results in greenhouse pollination of Habanero pepper in Mexico. It was revealed that there was no substantial difference in the quality of fruit produced by both the bee species but the number and percentage of fruit abortion was lesser in bumble bees than the native bee (Gerardo *et al.* 2008; Vossler *et al.* 2018). There are two studies that have reported on the pollination effectiveness of *Melipona quadrifasciata* for tomato grown in greenhouses in Brazil.

Bomfim *et al.* (2014), evaluated two stingless bee species in a greenhouse namely, *Melipona subnitida* Ducke and *Scaptotrigona* sp. during the flowering of seeded as well as seedless mini watermelon (*Citrullus lanatus* T.) varieties. They concluded that *M. subnitida* did not adapt to greenhouse mini watermelon cultivation under the experimental conditions. However, *Scaptotrigona* sp. adapted suitably as a pollinator in the greenhouse and actively collected floral resources soon after their introduction. Thus they can be used for mini watermelon pollination in protected culture.

In studies conducted by Samoskorn *et al.* (2010) for the efficacy of the stingless bee *Trigona pegdeni* to pollinate cucumbers, the fruit set rate was found to be lower than open pollination. It might be due to the fact that *Trigona pegdeni* stingless bees are smaller than *Lepidotrigona terminata* stingless bees. A remarkable increase in the yield attributes was observed such as fruit length, fruit girth, fruit weight, number of fruits per plant and yield/plant in pollinated crop over control plots at Tamil Nadu Agricultural University, Coimbatore and Srivilliputhur of southern India (Tej *et al.* 2017).

In Malaysia, the potential use of stingless bees as pollinator for greenhouse cucumber was studied using the stingless bees, *Heterotrigona itama* revealed heavier, longer and larger cucumbers compared to those produced from pollination without stingless bees. However, in terms of dried weight seed, firmness and colour of cucumbers, no significant difference between the treatments were detected (Azmi *et al.* 2017).

Fig 13.8. Stingless bee *Tetragonula iridipeniis* pollinating watermelon crop under net house.

Fig 13.9. Stingless bee *Lepidotrigona ventralis* pollinating tomato crop under greenhouse.

Pollination potential of stingless bee, *T. iridipennis* on ash gourd under protected conditions revealed higher fruit set, fruit quality parameters (fruit weight, length, seed number, seed weight and reduced misshapen fruits) was higher as compared to honeybees in open pollinated crop. Manyfold per cent increase was recorded in production and productivity parameters (Chauhan *et al.* 2019). In water melon, higher per cent fruit set and fruit quality reported by Chauhan and Singh (2021) by using stingless bees as pollinators against open pollination and pollinator exclusion (Fig 13.8). Chauhan *et al.* (2019) reported many fold increase in production and quality in cucumber, ash gourd and tomato (Fig 13.9).

Today, it is a known fact that the stingless bees have facilitated the passage of a new era to utilize them as an important pollinator. It was in 1999 that the first detailed review on the aspect of stingless bees in crop pollination appeared publically (Heard 1999). He reviewed findings of different authors and reported that stingless bees are effective and important pollinators of nine crops and that they contribute to pollination in 60 other species out of the 90 crop species they were found visiting as foragers. Over the past years, several new studies on stingless bee pollination has appeared from around the world. After the review by Heard (1999) there was a clear trend towards a more experimental approach using enclosures such as bags, cages and greenhouses (Biesmeijer and Slaa 2004; Nieh 2004). These inter-specific differences allow for selection of the most appropriate stingless bee for a given crop species and crop breeding system (greenhouse, open field, etc.). It can be concluded that commercial pollination with stingless bees are not established till today. Nevertheless, several biological features of stingless bee offer them as strong contenders for commercial pollination services partly because stingless bees are true generalists, collecting nectar and pollen from a vast array of plants (Heithaus 1979 a, b; Roubik 1989). Many stingless bee species have proven to forage suitably even in enclosed areas and under requisite climatic conditions. These bees can forage the whole year round. This indicate that they are especially appropriate for off-season production of crops in green houses. Most species of stingless bees have a foraging range shorter than that of the honeybees which may enhance foraging efficiency in confined spaces (Visscher and Seeley 1982; Seeley 1985; Katayama 1987; Kakutani *et al.* 1993). Stingless bees have many advantages such as they are generally less harmful to humans and other domesticated animals and they are able to forage effectively in glasshouses (Kakutani *et al.* 1993). Propagation of colonies contributes to preservation of biodiversity by conserving populations of species that may otherwise decline owing to human disruption of ecosystems (Inoue *et al.* 1984). Colonies are rarely able to abscond as the old queen is rather flightless (Inoue *et al.* 1984); and they are resistant to the diseases and parasites of honeybees (Delfinado-Baker *et al.* 1989). Stingless bees have yet not been commercially bred on a large scale for pollination purposes although they have been proven to possess great potential for the pollination of many crops including avocado, coconut, coffee, guava, and mango and rose apple grown in open culture as well as strawberry, sweet pepper and tomato grown under greenhouse conditions.

■

CHAPTER 14

Stingless Bees – Present and Future in India

Stingless bees

Like all other bees, stingless bees are members of the class Insecta, order Hymenoptera and family Apidae, tribe Meliponini, sub tribe Meliponina and are closely related to common honeybees and bumble bees. Stingless bees play a very important role in the primary health care through production of medicinal hive products such as honey, propolis and bee bread. Like honeybees, they live in cavities with a queen, workers (sterile females) and drones (males). Stingless bees consist of a large group of bees having about 500 described species found throughout the tropical and subtropical regions of the globe. Meliponines have stingers but they are significantly reduced and cannot be used for defence and thus the term "stingless" is being used to designate the species. Some species have sufficiently strong mandibles to inflict a mild bite, pull hairs or even crawl into ears and nostrils of the intruders. Others emit a caustic liquid from the mouth which on contact with the skin causes intense irritation.

In India, stingless bee keeping is practiced in North Eastern States (Nagaland, Assam, Anurachal Pradesh, Manipur, Meghalaya, Tripura, Mizoram and Sikkim), Southern states (Kerala, Tamil Nadu, Andhra Pradesh and Karnataka), Eastern states (West Bengal and Odisha) and in pockets of Punjab and Gujarat. North Eastern states are an important biodiversity hotspot having vast bee flora and fauna. Meliponiculture have been practiced in all the above-mentioned states since immerorial. Presently, all these states practice beekeeping in traditional manner (in simple, tree trunks, logs, bamboos and underground domiciles) with low honey production and more bee colony loses.

In Kerala, Karnataka and Tamil Nadu, more focus was inclined towards the biology of stingless bees and their taxonomy while most of the work regarding the scientific stingless bee keeping was studied in the North Eastern India. Both Nagaland and Assam states began scientific works on stingless bees almost at the same time but Nagaland having the largest diversity of stingless bees, developed the scientific hives. The beekeepers are shifting from traditional beekeeping to scientific beekeeping because of their twin role in increasing crop productivity through pollination and honey production.

Stingless bees are one of the most promising pollinators of different crops for the following reasons: they are harmless to beekeepers and greenhouse workers, visit a wide range of crops (polylecty), are tolerant to high temperatures, are active throughout the year and can be transported easily, not succeptible to diseases and have less enemies.

Requirement of scientific stingless bee keeping

The NER region of India is producing 40.51 L million tons of vegetables from 3.68 lakh/ha with productivity level of 11 tons/ha which is slightly lower as compared to national average (17.60 tons/ha). To increase the production and productivity, bee pollination is proven to play an important role. As the agriculture is shifting towards protected cultivation, stingless bees can be proved as important pollinators of different crops grown under protected conditions hence can be utilized commercially for delivering pollination in order to employ stingless bees in pollination services and to increase honey production, scientific beekeeping and pollination management strategies needs to be formulated.

Importance of stingless bees

Stingless bees are ecologically important species of bees. They provide different high value hive products such as honey, wax, propolis and bee brood. Apart from those, they provide pollination services to different crops that results in increased productivity and production of different crops. They also maintain the ecological biodiversity by rendering pollination services.

Current scenario

Stingless bee domicilation

Presently, stingless bees are kept in log hives, wooden hives and underground in small buckets, flower pots etc. New hives have been designed and are under trials for the successful domicilation of stingless bees to increase the honey production without disturbing the brood. Different splitting methods of colony division are tested and used for successful splitting and raising of daughter colonies.

Diversity of stingless bees in Nagaland

Nagaland is bestowed with huge flora and fauna biodiversity and stingless bees are present in every pocket of Nagaland thus making this place a special stingless bee hub. Out of six species of stingless bees reported from North East Indian states, all of these six species are available in Nagaland. The six different species, viz. *Tetragonula iridipennis*, *Lophotrigona canifrons*, *Lepidotrigona ventralis, Lepidotrigona arcifera, Tetragonula gressitti* and *Tetragonula laeviceps* are identified in Nagaland province of India. There is every possibility that many more species are yet to be discovered as it is one of the biodiversity hotspots.

Scope of stingless bees in agriculture

In North east region of India, most agro-ecosystems are organic which directly increase the scope of Meliponiculture. Most of the farmers are keeping stingless bees in the traditional manner for honey extraction but are still unaware of their pollination potential due to lack of studies in this field. Farmers grow seasonal vegetables and fruits but with changing scenario, the agriculture is getting intensified and is done under protected conditions. Under open conditions, natural pollinators are available but pollination under protected conditions, the avenues are still unexplored. With stingless bees as the native pollinator, opportunities are there to develop it into a commercial pollinator of different crops grown under protected conditions. Moreover, *Apis mellifera* which is used as a commercial pollinator in different parts of the globe is not present in Nagaland, thus providing a positive step in favour of stingless bees to be developed as commercial pollinator under NER regions of India. In litchi, cucumber and tomato, comparative pollination potential of honeybees and stingless bees were studied which disclosed the potential of stingless bees in these crops as the productivity was increased manifold when compared to those without pollination.

Constraints in stingless bee rearing

The number of stingless bee colonies are increasing gradually, but the average production of honey from the colonies is not increasing because of old traditional practices and hives used by beekeepers. Some of the constraints hampering the development and growth of stingless bee culture are:

- Stingless bee diversity is still not fully explored.
- Unavailability of scientific techniques in stingless bee rearing.
- Lack of infrastructure and unavailability of superior domiciles.
- Technical expertise for colony management to increase honey production.
- Lack of awareness to increase the crop productivity through managed pollination.
- R&D efforts in research for pest management.
- Loss of natural fauna supporting bee population.
- Consumer awareness on benefits of honey and other honeybee products for nutrition and health.

References

Abak K, Dasgan H Y, Ikiz O, Uygun N, Sayalan M, Kaftanoglu O and Yeninar. 1997. Pollen production and quality of pepper grown in unheated greenhouses during winter and the effects of bumble bees (*Bombus terrestris*) pollination on fruit yield and quality. *Acta Horticulturae* 437: 303–307.

Abak K, Ozodogan A O, Dasgan H Y and Derin K. 1999. Effectiveness of bumble bees as pollinator for egg plants grown in unheated greenhouse. *Acta Horticulturae* 422: 197-203.

Abak K, Sari N and Paksoy M. 1995. Effciency of bumble bees on the yield and quality of eggplant and tomato grown in unheated glasshouses. *Acta Horticulturae* 412: 268–274.

Absy M, Rech A and Ferreira M. 2018. *Pollen Collected by Stingless Bees: A Contribution to Understanding Amazonian Biodiversity*. (In) *Pot-Pollen in Stingless Bee Melittology*, pp 29-46. AG Press.

Agashe S N and Rangaswamy B E. 1997. Melittopalynological studies of honey samples from Bagardka, Dakshin Kannad district, Karnataka. *Indian Bee Journal* 59: 8-11.

Aguilar I and Briceño D. 2002. Sounds in *M. costaricensis* (Apidae: Meliponini): effect of sugar concentration and nectar source distance. *Apidologie* 33: 375–388.

Aguilar I and Sommeijer M. 2001. The deposition of anal excretions by *Melipona favosa* foragers (Apidae: Meliponinae): behavioural observations concerning the location of food sources. *Apidologie* 32: 37–48.

Aguilar I and Sommeijer M. 1996. Communication in stingless bees: are the anal substances deposited by *Melipona favosa* scent marks? Proceedings of Experimental and Applied Entomology. *Netherlands Entomological Society* 7: 57–63.

Ahn C K, Choe Y C Um, Cho U C and Park J C. 1988. Effects of the pollination and growth regulators treatment on preventing the malformation and accelerating the growth of strawberry fruit. *Research Reports of the Rural Development Administartion Horticultural* 30(3): 22-30.

Akaratanakul P. 1987. *Honeybee diseases and enemies in Asia a practical guide*. FAO, Rome.

Allard H A. 1911. Some experimental observations concerning the behaviour of various bees in their visits to cotton blossoms. *American Naturalist* 45: 607-685.

Almeida da Silva I A, Sarmento da Silva T M, Camara C A, Queiroz N, Magnani M, Santos de Novais J, Soledade L E B, Edeltrudes de Oliveira L, Lucia de Souza A and Gouveia de Souza A. 2013. Phenolic profile, antioxidant activity and palynological analysis of stingless bee honey from Amazonas, Northern Brazil. *Food Chemistry* 141: 3552–3558.

Alvarenga P E F. 2008. Levantamento da fauna de abelhas (Hymenoptera, Apidae, Meliponina) na mata Santa Tereza, estação ecológica de Ribeirão Preto-SP, e limitantes da densidade de seus ninhos. Master thesis, FFCLRP-USP.

Alves D J. 2006. Uso do marcador molecular RAPD para estudo de polimorfismos em populacoes de *Tetragonisca angustula* L. (Apidae: Meliponinae). Dissertacao de Mestrado em Genetica e Melhoramento. Universidade Estadual de Maringa, Maringa, PR. 45.

Alves R, Justina G, Souza B D A, Dias C and Sodre G D S. 2006. Criação de abelhas nativas sem ferrao (Hymenoptera: Apidae): Autosustentabilidade na comunidade de Joia do Rio, município de Camaçari, estadoda Bahia. Magistra.

Anderson K E, Sheehan T H, Eckholm B J, Mott B M and DeGrandi-Hoffman G. 2011. An emerging paradigm of colony health: microbial balance of the honeybee and hive (*Apis mellifera*). *Insectes Sociaux* 58: 431-444.

Anonymous. 1997. Ethics and equity in conservation and use of genetic resources for sustainable food security: proceedings of a workshop to develop guidelines for the CGIAR, 21-25 April 1997, Foz do Iguaçu, Brazil

Anonymous. 2019. *Biennial report AICRP Honeybees and Pollinators*. pp 1-73

Anonymous. 2019. *Report of impact of neonicotenoids on pollinators*. Presented in Review meeting of Impact of Neonicotenoids on pollinators, Division of Entomology, IARI, New Delhi.

Antonini Y, Soares S M and Martins R P. 2006. Pollen and nectar harvesting by the stingless bee *Melipona quadrifasciata anthidioides* (Apidae: Meliponini) in an urban forest fragment in Southeastern Brazil. *Neotropical Fauna and Environment* 41: 209–215.

Antunes O T, Calvete E O, Rocha H C, Nienow A A, Cecchetti D, Riva E and Maran R E. 2007. Produção de cultivares de morangueiro polinizadas pela abelha jataí em ambiente protegido. *Horticultura Brasileira* 25: 94-99.

Araujo V A, Antonini Y and Araújo A P A. 2006. Diversity of bees and their floral resources at altitudinal areas in the southern Espinhaco range, Minas Gerais, Brazil. Ecology, Behavior and Bionomics. *Neotropical Entomology* 35(1): 30-40.

Asada S and Ono M. 1996. Crop pollination by Japanese bumble bee (*Bombus* spp.). Tomato foraging behaviour and pollination efficiency. *Applied Entomological Zoology*.

Iwasaki M. 1995. Introduction of commercial bumble bees into Japan. *Honeybee Science* 16: 17-23.

Asada S and Ono M. 1997. Tomato pollination with Japanese native bumblebees (*Bombus* spp.). *Acta Horticulturae* 437: 289–292.

Atkins E L, Kellum D and Atkins K W. 1981. Reducing pesticide hazards to honeybees: mortality prediction techniques and integrated management strategies. Science Leaflet 2883. Division of Agricultural Sciences, University of California, Davis, CA, USA.

Attal Y Z A, Kasrawi M A and Nazer I K. 2003. Influence of bumble bee technique on greenhouse tomato production. *Agricultural and Marine Sciences* 8(1): 21-26.

Azmi W A, Samsuri, Hatta N M M, Ghazia R and Seng C T. 2017. Effects of stingless bee (*Heterotrigona itama*) pollination on greenhouse cucumber (*Cucumis sativus*). *Malaysian Applied Biology* 46(1): 51–55.

Baker D M and Baker E W. 1988. New mites from the nests of stingless bees (Apidae: Meliponinae) from Asia. *International Journal of Acarology* 14: 131-139.

Baker E W and Delfinado M. 1985. An unusual species of *Neocypholaelaps* (Acari: Amerosiidae) from Asia. *International Journal of Acarology* 14: 127-136.

Banda H E J and Paxton R J. 1991. Pollinaton of greenhouse tomato by bees. *Acta Horticulturae* 288: 194-196.

Bankova V. 2009. Chemical diversity of propolis makes it a valuable source of new biologically active compounds. *Journal of ApiProduct ApiMed Science* 1: 23-28.

Baptistella A R, Souza C C M, Santana W C and Soares A E E. 2012. Techniques for the *in vitro* production of queens in stingless bees (Apidae, Meliponini). *Sociobiology* 59: 297–310.

Barbosa F M, Alves R M O, Souza B A and Carvalho C A L. 2013. Nest architecture of the stingless bee *Geotrigona subterranea* (Friese 1901) (Hymenoptera: Apidae: Meliponini). *Biota Neotrop* 13(1): 147-152.

Barbosa W F, Smagghe G and Guedes R N C. 2015. Perspective Pesticides and reduced-risk insecticides, native bees and pantropical stingless bees: pitfalls and perspectives. *Pest Management Science* 71: 1049–1053.

Barbosa W F, Tome H V V, Bernardes R C, Siqueira M A L, Smagghe G and Guedes R N C. 2015b. Biopesticide-induced behavioral and morphological alterations in the stingless bee *Melipona quadrifasciata*. *Environment Toxicology Chemistry* 34: 2149–2158.

Barbosa W F, De Meyer L, Guedes R N C and Smagghe G. 2015c. Lethal and sublethal effects of azadirachtin on the bumblebee *Bombus terrestris* (Hymenoptera: Aapidae). *Ecotoxicology* 24: 130–142.

Bartelli B F, Santos A O R and Ferriera N F H. 2014. Colony performance of *Melipona quadrifasciata* (Hymenoptera, Meliponina) in a greenhouse of *Lycopersicon esculentum* (Solanaceae). *Sociobiology* 61(1): 60-67.

Barth F G, Hrncir M and Jarau S. 2008. Signals and cues in the recruitment behaviour of stingless bees (Meliponini). *Journal of Comparative Physiology A*. 194: 313-327. DoI: 10.1007/s00359-008-0321-7

Barth O M and Luz C F P. 1998. Melissopalynological data obtained from a mangrove area near to Rio de Janeiro, Brazil. *Journal of Apicultural Research* 37(2): 155–163.

Barth O M. 2013. *Palynology serving the stingless bees*. (In) A Legacy of Stingless Bees. pp. 285-294. DoI 10.1007/978-1-4614-4960-7_20.

Basrawi F, Ahmad A H, Idris D M N D, Maarof M R M, Chand M and Ramli A S. 2017. *Engineering economic analysis of meliponiculture in Malaysia considering current market price*. (In) MATEC Web of Conferences. EDP Sciences.

Batista M A, Ramalho M and Soares A E E. 2003. Nesting sites and abundance of Meliponini (Hymenoptera: pidae) in heterogeneous habitats of the Atlantic rain forest, Bahia, Brazil. *Lundiana* 4: 19-23.

Batra S W T. 1995. Bees and pollination in our changing environment. *Apidologie* 26: 361-370.

Behere G T, Tay W T, Russell D A, Heckel D G, Appleton B R, Kranthi K R and Batterham P. 2007. Mitochondrial DNA analysis of field populations of *Helicoverpa armigera* (Lepidoptera: Noctuidae) and of its relationship to *H. zea*. *BMC Evolutionary Biology* 7: 117.

Berger L A, Vassiere B E, Moffet O J and Merritt J. 1988. *Bombus* spp. (Hymenoptera: Apidae) as pollinators of male-sterile upland cotton on the Texas high plains. *Environmental Entomology* 17(5): 789-794.

Bhambura.1958. Further studies on the importance of honeybees in pollination of cucurbitaceae. *Indian Bee Journal* 20: 189-191.

Bhargava H R, Jyothi J V A, Bhushanam M and Surendra N S. 2009. Pollen analysis of *Apis* honey, Karnataka, India. *Apiacta* 44: 14- 19.

Biesmeijer J C, Born M, Lukacs S and Sommeijer M J. 1999a. The response of the stingless bee *Melipona beecheii* to experimental pollen stress, worker loss and different levels of information input. *Journal of Apicultural Research* 38: 33–41.

Biesmeijer J C and de Vries H. 2001. Exploration and exploitation of food sources by social insect colonies: a revision of the scout-recruit concept. *Behavioural Ecology and Sociobiology* 49: 89–99.

Biesmeijer J C and Ermers M C W. 1999. Social foraging in stingless bees: how colonies of *Melipona fasciata* choose among nectar sources. *Behavioural Ecology and Sociobiology* 46: 129–140.

Biesmeijer J C and Slaa E J. 2004. Information flow and organization of stingless bee foraging. *Apidologie* 35: 143–157.

Biesmeijer J C, Giurfa M, Koedam D, Potts S G, Joel D M and Dafni A. 2005. Convergent evolution: Floral guides, stingless bee nest entrances, and insectivorous pitchers. *Natur wissen schaften* 92: 444-450.

Biesmeijer J C, van Nieuwstadt M G L, Lukács S and Sommeijer M J. 1998.The role of internal and external information in foraging decisions of *Melipona* workers (Hymenoptera: Meliponinae). *Behavioral Ecology and Sociobiology* 42(2): 107–116.

Bijlsma L, De Bruijn L L M, Martens E P and Sommeijer M J. 2006. Water content of stingless bee honeys (Apidae, Meliponini): interspecific variation and comparison with honey of *Apis mellifera*. *Apidologie* 37: 480-486.

Bingham C T. 1897. *Hymenoptera, Wasp and Bees*. (In) WT Blandford (Ed). *The fauna of British India including Ceylon and Burma*. Taylor and Francis, London, pp 579-1897.

Blum M S, Crewe R M, Kerr W E, Keith L H, Garrison A W and Walker M M. 1970. Citral in stingless bees: isolation and functions in traillaying and robbing. *Journal of Insect Physiology* 16: 1637– 1648.

Bohart G E and Nye W P. 1960. Insect pollinators of carrots in Utah. *Bullettin of Utah Agricultural Experiment Station* 419: 16-17.

Bohart G E. 1972. Management of wild bees for the pollination of crops. *Annual Review of Entomology* 17: 287-312.

Bomfim I G A, Bezerra A D M, Nunes A C, Aragao F A S and Freitas B M. 2014. Adaptive and foraging behavior of two stingless bee species (Apidae: Meliponini) in greenhouse mini watermelon pollination. *Sociobiology* 61: 502-509.

Boogert N J, Hofstede F E and Aguilar Monge I. 2006. The use of food source scent marks by the stingless bee *Trigona corvina* (Hymenoptera: Apidae): the importance of the depositor's identity. *Apidologie* 37: 366-375.

Borsato D M, Prudente A S, Boscardin P M, Borsato A V, Luz C F P, Beatriz H L, Maia N S, Cabrini D A, Otuki M F, Miguel M D, Farago P V and Miguel O

G. 2014. Topical anti-inflammatory activity of a monofloral honey of *Mimosa scabrella* provided by *Melipona marginata* during winter in southern Brazil. *Journal of Medicinal Food* 17(7): 817–825.

Bradbear N. 2009. Bees and their role in forest livelihoods A guide to the services provided by bees and the sustainable harvesting, processing and marketing of their products. FAO, Rome. *Non-Wood Forest Products* 19: 1–194.

Breed M D, Cook C N and Krasnec M O. 2012. Cleptobiosis in social insects. *Psyche- a Journal of Entomology* 7(2): 1-7.

Bruijn L L M de and Sommeijer M J. 1997. Colony foraging in different species of stingless bees (Apidae, Meliponinae) and the regulation of individual nectar foraging. *Insectes Sociaux* 44(1): 35–47.

Calzoni G L and Speranza A. 1996. Pear and plum pollination: honeybees, bumble bees or both? *Acta Horticulturae* 423: 83–90.

Camacho E. 1966. Da˜no que las abejas jicotes del g´enero Trigona causan a los ´arboles de Macadamia. *Turrialba* 16: 193–194.

Camargo J M F and Pedro S R M. 2003. Meliponini neotropicais: o gênero *Partamona* Schwarz, 1939 (Hymenoptera, Apidae, Apinae) - bionomia e biogeografia. *Revista Brasileira de Entomologia*. 47: 311-372. http://dx.DoI.org/.

Camargo J M F and Pedro S R M. 2004. Meliponini neotropicais: o gênero *Ptilotrigona* Moure (Hymenoptera, Apidae, Apinae). *Revista Brasileira de Entomologia* 48: 353-377.

Camargo J M F and Roubik DW. 1991. Systematics and bionomics of the apoid obligate necrophages: the *Trigona hypogea* group (Hymenoptera: Apidae; Meliponinae). *Biological Journal of Linneaous Society* 44: 13– 40.

Campos L A O. 1987. Abelhas indígenas sem ferrão: o que são? *Information Agro* 13(147): 42-44.

Can-Alonso C, Quezada-Euan J J G, Xiu-Ancona P, Moo-Valle H, Valdovinos-Nunez G R and Medina-Peralta S. 2005. Pollination of 'criollo' avocados (*Persea americana*) and the behaviour of associated bees in subtropical Mexico. *Journal of Apicultural Research* 44: 3-8.

Cane J and Sipes S. 2006. *Characterizing floral specialization by bees: Analytical methods and a revised lexicon for oligolecty*. (In) Waser N M, Ollerton J (Eds.). *Plant-pollinator interactions from specialization to generalization*. University of Chicago Press, Chicago, pp 99–122.

Carvalho C A L, Moreti A C C C, Marchini L C, Alves R M O and Oliveira P C F. 2001. Pollen spectrum of honey of 'Uruçu' bee (*Melipona scutellaris* Latreille, 1811). *Revista Brasileira de Biologia* 61: 63–67. DoI:10.1590/S0034-71082001000100009

Casida J E and Durkin K A. 2013. Neuroactive insecticides: targets, selectivity, resistance, and secondary effects. *Annual Review of Entomology* 58: 99–117.

Cauich O, Quezada-Eubn J J G, Macias-Macias J O, Reyes-Oregel V, Medina-Peralta S and Parra-Tabla V. 2004. Behavior and pollination efficiency of *Nannotrigona perilampoides* (Hymenoptera: Meliponini) on greenhouse tomatoes (*Lycopersicum esculentum*) in Subtropical Mexico. *Horticultural Entomology* 97: 475-481.

Cavelkante V M, Oleviera V P T and Cruz-Landim C. 2000. Comparative study of wax glands in four Meliponini bees (Hymenoptera, Apidae) producing different quantities of wax. *Iheringia Séri Zoology* 89: 193-198.

Chaudhary S and Gupta J K. 2000. Influence of worker brood survival on the morphoiogical, behavioural and economical characters of *Apis mellifera*. *Indian Bee Journal* 62: 55-58.

Chauhan A. 2011. *Refinement of bumble bee rearing technology and its use in cucumber pollination*. MSc thesis, Dr Y S Parmar University of Horticulture and Forestry, Nauni, Solan, India.

Chauhan A. 2015. *Studies on pests and diseases of bumble bee, Bombus haemorrhoidalis Smith*. PhD thesis, Dr Y S Parmar University of Horticulture and Forestry, Nauni, Solan, India.

Chauhan A. 2018. *Stingless bees in Nagaland*. Presented in Indo-Australian Workshop on stingless bees. November 2019, New Delhi.

Chauhan A and Singh H K. 2019. Nest architecture studies of stingless bees, *Tetragonula iridipennis* and *Lepidotrigona ventralis* in Nagaland. *Indian Journal of Entomology* 81(3): 614-617

Chauhan A and Singh H K. 2020. *Beekeeping in Nagaland- a training manual*. AICRP on Honeybees and Pollinators, Medziphema, Nagaland.

Chauhan A and Singh H K. 2021. Mapping of stingless bee fauna in Nagaland. *Bioscan* 16(1): 1-4.

Chauhan A, Rana B S and Shruti Kumari. 2015. Karyological studies on bumble bee, *Bombus haemorrhoidalis* Smith in Himachal Pradesh. *International Journal of Farm Sciences* 5(2): 129-139.

Chauhan A, Katna S and Rana B S. 2013. Ethological studies on bumble bees under subtropical conditions in India. *Journal of Medical Science and Clinical Research* 1(3): 125-129.

Chauhan A, Katna S, Rana B S and Rana Kiran. 2013. Morphometric studies on bumble bee, *Bombus rufofasciatus* Smith in Himachal Pradesh. *Bioscan* 11(2): 721-723.

Chauhan A, Katna S and Rana B S. 2013. Life cycle of bumble bee *Bombus haemorrhoidalis* Smith in Himachal Pradesh. *Insect Environment* 19(3): 183-186.

Chauhan A, Katna S, Rana B S and Miyan H V. 2013. Field establishment of artificially reared bumble bee (*Bombus haemorrhoidalis* Smith) colonies in Himachal Pradesh. *Insect Environment* 19(3): 159-161.

Chauhan A, Rana B S and Katna S. 2015. Different castes in bumble bee, *Bombus haemorrhoidalis* Smith. *Ecology, Environment and Conservation* 22(1): 297-306.

Chauhan A, Rana B S, Sharma K C and Katna S. 2016. Seasonal incidence and biology of conopid fly, *Physocephala tibialis* in bumble bee, *Bombus haemorrhoidalis* Smith in India. *Ecology, Environment and Conservation* 22(2): 819-823.

Chauhan A, Rana B S, Thakur M and Katna S. 2017. RAPD-PCR studies of bumble bee (*Bombus haemorrhoidalis* and *Bombus rufofasciatus*) and their important pests (*Physocephala tibialis* and *Aphomyia sociella*). *Ecology, Environment and Conservation* 23(4): 364-368.

Chauhan A, Singh H K, Kumaranag K M and Chakrabarty P K. 2019. *Stingless bee rearing and its potential in pollination of vegetable crops under protected conditions*. Booklet published by Department of Entomology, SASRD,

Nagaland University, Medziphema,11 p.

Chauhan A, Singh H K and Awomi S O. 2020. *Impact of stingless bee Tetragonula iridipennis pollination in watermelon Citrullus lanatus under protected conditions*. Abstract Book, Indian Horticulture Summit, 14-16 February, 2020 Chitrakoot Madhya Pradesh. p. 171.

Chauhan A, Singh H K and Kumaranag K M. 2019. Impact of stingless bee, *Tetragonula iridipennis* pollination in ash gourd. *Indian Journal of Entomology* 81(4): 854-863.

Chauhan Avinash and Thakur R K. 2014. Studies on nest architecture and pollination potential of bumble bee (*Bombus haemorrhoidalis*). *Indian Journal of Ecology* 41(1):158-164.

Chauhan Avinash, Rana B S and Katna S. 2014. *Bumble bees in India- their rearing and management*. Lambert's Germany.

Chauhan Avinash, Rana B S and Katna S. 2014. Successful rearing of bumble bee, *Bombus haemorrhoidalis* Smith year-round in Himachal Pradesh in India. *International Journal of Current Research* 6(12): 10891-10896.

Chittka L and Leadbeater E. 2005. Social learning: public information in insects. *Current Biology* 15(21): 703-713. R869-871. DoI: 10.1016/j.cub.2005.10.018

Ciar R R, Bonto L S, Bayer M H, Rabajante J F, Lubag S P, Fajardo A C and Cervancia C R. 2013. Foraging behaviour of stingless bee (*Tetragonula* sp), distance, direction and height of preferred food source. arXiv:1310.3919 [q-bio.PE].

Conama. 2004. Resolucao Conama nº 346, de 16 de agosto de 2004 Publicada no DOU no 158, de 17 de agosto de 2004, Seção 1, página 70, Gestão de espécies – Fauna, [online] www.areaseg.com/conama/ 2004/346-2004.pdf.

Constance S S and Francis A D. 2001. *Bombus impatiens* (Hymenoptera: Apidae): an alternative to *Apis mellifera* (Hymenoptera: Apidae) for lowbush blue berry pollination. *Journal of Economic Entomology* 94(3): 609-616.

Contrera F A L, Menezes C and Venturieri G C. 2011. New horizons on stingless bee keeping (Apidae, Meliponini). *Revista Brasileira de Zootecnia* 40: 48-51.

Corbet S A, Beament J W L and Erisikowitch D. 1982. Are electrostatic forces involved in pollen transfer. *Plant cell and Environment* 5: 125-129.

Corbet S A, Chapman H and Saville N. 1988. Vibratory pollen collection and flower form: bumble bee of *Actinidia*, *Borago* and *Polygonatum*. *Functional Ecology* 2: 147-155.

Corbet S A, Williams I H and Osborne J L. 1991. Bees and the pollination of crops and wild-flowers in the European Community. *Bee World* 72: 47-59.

Corbett S A and Willmer P G. 1980. Pollination of the yellow passion fruit: nectar, pollen and carpenter bees. *Journal of Agricultural Science* 95: 655–666.

Cortopassi-Laurino M. 1979. Observacoes sobre atividades de machos de *Plebeia droryana* Friese (Apidae, Meliponinae). *Revista Brasilia de Entomologica* 24: 177–191.

Cortopassi-Laurino M and Ramalho M. 1988. Pollen harvest by Africanized *Apis mellifera* and *Trigona spinipes* in Sao Paulo: botanical and ecological views. *Apidologie* 19: 1-24.

Cortopassi-Laurino M, Imperatriz-Fonseca V L, Roubik D W, Dollin A, Heard T, Aguilar I, Venturiei G C, Eardley C and Nogueira-Neto P. 2006. Global meliponiculture: challenges and opportunities. *Apidologie* 37(2): 275-292.

Cortopassi-Laurino M, Knoll F R N, Ribeiro M F, Heemert C and Ruijter A. 1991. Food plant preferences of *Friesella schrottkyi*. *Acta Horticulture* 288: 382–85.

Cortopassi-Laurino M and Nogueira-Neto P. 2016. Abelhas sem ferrao do Brasil. São Paulo: EDUSP. 124 p.

Costa L M, Grella T C, Barbosa R A, Malaspina O and Nocelli R C F. 2015. Determination of acute lethal doses (LD50 and LC50) of imidacloprid for the native bee *Melipona scutellaris* Latreille, 1811 (Hymenoptera: Apidae). *Sociobiology* 62(4): 578- 582.

Costa R A C and Da Cruz-Landim C. 2005. Hydrolases in the hypopharyngeal glands of workers of *Scaptotrigona postica* and *Apis mellifera* (Hymenoptera, Apinae). *Genetetics Molecular Research* 4: 616-623.

Cox D W. 1999. Disorders of copper transport. *Bullettin of Br Medicine* 55: 544–555.

Craig C L, Webber R S and Bernard G D. 1996. Evolution of predator prey systems: Spider foraging plasticity in response to the visual ecology of prey. *Journal of Kansas Entomological Society* 69: 132-134.

Crane E. 1999. The past and present status of bee keeping with stingless bees. *Bee World* 73: 29-42.

Crane E. 1991. *Apis* species of tropical Asia as pollinators and some rearing methods for them. *Acta Horticulturae* 288: 29-48.

Crane E. 1992. The past and present status of beekeeping with stingless bees. *Bee World* 73: 29-42.

Cruz D de O, Freitas B M, Silva L A, da Silva S E M D and Bomfim I G A. 2005. Pollination efficiency of the stingless bee *Melipona subnitida* on greenhouse sweet pepper. *Pesquisa Agropecuaria Brasileira Brasilia* 40: 1197-1201.

Cruz-Landim C D, Moraes Regina L M S D, Salles Heliana C and Reginato Rejane D. 1998. Note on glands present in Meliponinae (Hymenoptera, Apidae) bee legs. *Revista Brasileira de Biologica* 15: 159–165.

Cruz-Landim C. 1967. Estudos comparativos de algumas glândulas das abelhas (Hymenoptera, Apoidea) e respectivas implicações evolutivas. *Arq. Zool. Est. S. Paulo, São Paulo* 15(3): 177-290.

Cufford P T P. 1973. Increasing bumble bee densities in red clover seed production areas. *New Zealand Journal of Experimental Agriculture* 1(4): 377-399.

Cunningham J P, Hereward J P, Heard T A, De Barro P J and West S A. 2014. Bees at war: interspecific battles and nest usurpation in stingless bees. *American Naturalist* 184: 777–786.

Cutler G C, Scott-Dupree C D, Sultan M, McFarlane A D and Brewer L. 2014. A large-scale field study examining effects of exposure to clothianidin seed-treated canola on honeybee colony health, development, and overwintering success. *Peer Journal* 2: 652.

Dag A and Kammer Y. 2001. Comparison between the effectiveness of honeybee (*Apis mellifera*) and bumble bee (*Bombus terrestris*) as pollinators of greenhouse sweet pepper (*Capsicum annuum*). *American Bee Journal* 141: 447-448.

Dall S R, Giraldeau L A, Olsson O, McNamara J M and Stephens D W. 2005. Information and its use by animals in evolutionary ecology. *Trends Ecology and Evolution* 20: 187-193. DoI: 10.1016/j.tree.2005.01.010

Danaraddi C S and Viraktamath S. 2009. Morphometrical studies on the stingless bee, *Trigona iridipennis* Smith. *Karnataka Journal of Agricultural Sciences* 22: 796-797.

Danaraddi C S, Hakkalappanavar S, Biradar S B, Tattimani M and Vinod S K. 2011. Studies on foraging behaviour of stingless bee, *Trigona iridipennis* Smith at Dharwad, Karnataka. *International Journal of Forestry and Forestry Crop Improvement* 2(2): 163-169.

Danchin E, Giraldeau L A, Valone T J and Wagner R H. 2004. Public information: from nosy neighbors to cultural evolution. *Science* 305: 487-491.

Daners G and Telleria M C. 1998. Native vs. introduced bee flora: A palynological survey of honeys from Uruguay. *Journal of Apicultural Research* 37(4): 221-229.

Danforth B N. 1999. Emergence dynamics and bet hedging in a desert bee, *Perdita portalis*. *Proceedings of Royal Society of London B*. 266: 1985–1994.

Dasgan H Y and Ozdogan A O. 2004. Effectiveness of bumble bee pollination in anti-frost heated tomato greenhouses in the Mediterranean basin. *Turkey Journal of Agricultural Research* 28: 73-82.

Dasgan H Y, Ozdogan A O and Abak K. 1999. Comparison of honeybees and bumble bees as pollinators for melon (*Cucumis melo* L.) grown in greenhouses. *Acta Horticulturae* 492: 131-134.

Dayal Kiran and Rana B S. 2004. Record of domestication of *Bombus* species (Hymenoptera: Apidae) in India. *Insect Environment* 10(2): 64-65.

Marco de R J and Farina W M. 2003. Trophallaxis in forager honeybees (*Apis mellifera*): resource uncertainty enhances begging contacts? *Journal of Comparative Physiology A*. 189: 125–134.

De Sousa J M, de Souza E L, Marques G, Meireles B, de Magalhães Cordeiro A T, Gullón B and Magnani M. 2016. Polyphenolic profile and antioxidant and antibacterial activities of monofloral honeys produced by Meliponini in the Brazilian semiarid region. *Food Research International* 84: 61–68.

De Grandi-Hoffman G, Hoopingarner R and Klomparens K. 1986. Influence of honeybee (Hymenoptera: Apidae) in-hive pollen transfer on cross-pollination and fruit set in apple. *Environmental Entomology* 15: 723–725.

Dejtisakdi W, Areelop W, Masuthon S, Santisuk T, Wongsiri S and Juntawong N. 2005. *Melliferous plants for the stingless bees (Trigona collina Smith) in deciduous forest of Queen Sirikit Botanic Garden , Mae Rim , Chiang Mai Province*. In Proceedings of 43rd Kasetsart University Annual Conference, Thailand, 1-4 February, 2005. Subject: Plants 2005, pp 280–287.

Delfinado-Baker M, Baker E W and Phoon A C G. 1989. Mites (Acari) associated with bees (Apidae) in Asia, with description of a new species. *American Bee Journal* 129: 609–613.

Delplane Keith S. 1995. Bumble beekeeping: queen starter box. *American Bee Journal* 133: 743-745.

Devanesan S, Nisha M M, Bennet R and Shailaja K K. 2002. Foraging behaviour of stingless bee, *Trigona iridipennis* Smith. *Insect-Environment* 8(3): 131-133.

Devanesan S, Shailaja K K and Premila K S. 2009. *Status paper on stingless bee Trigona iridipennis* Smith. All India Co-ordinated Research Project on Honeybees and Pollinators (ICAR). Regional Agricultural Research Station, College of Agriculture, Vellayani, pp 45-61.

Dias B S F, Raw A and Imperatriz F V L. 1999. *International pollinators initiative: the Sao Paulo declaration on pollinators*. Report on the recommendations of the workshop on the conservation and sustainable use of pollinators in agriculture with emphasis on bees. Brazilian Ministry of the Environment, Brasília, p 79.

Dixon C V. 1987. *Beekeeping in Southern Mexico*. Conference of Latin Americanist Geographers Year Book. 13: 66-71. https://www.jstor.org/stable/25765682.

Dollin A E, Dollin L J and Sakagami S F. 1997. Australian stingless bees of the genus *Trigona* (Hymenoptera: Apidae). *Invertebrate Taxonomy* 11: 861-896.

Dorneles A L, de Souza Rosa A and Blochtein B. 2017. Toxicity of organophosphorus pesticides to the stingless bees, *Scaptotrigona bipunctata* and *Tetragonisca fiebrigi*. *Apidologie* 48: 612–620. DoI.org/10.1007/s13592-017-0502-x.

Downs S G and Ratnieks F L W. 2000. Adaptive shifts in honeybee (*Apis mellifera* L.) guarding behavior support predictions of the acceptance threshold model. *Behavioural Ecology* 11: 326–333.

Dreyling L. 1903. Über die wachsbereitenden Organe der Honigbiene. *Zool. Anz., Leipzig*. 26: 710-715.

Duarte A W S, Santos Vasconcelos M R, Domarques de Menezes A P, Chagas da Silva S, Oda-Souza M and Queijeiro López A M. 2012. Composition and antioxidant activity of honey from Africanized and stingless bees in Alagoas (Brazil): a multivariate analysis. *Journal of Apicultural Research* 51(1): 23-35.

Dutta K, Verma P C and Chatterjee A. 2008. Mapping of key bee flora of upper Gangetic region of India – a palynological assessment through regional honeys. *Grana* 47(2): 159-170. DoI: 10.1080/00173130801980109

Dyer F C. 2002. The biology of the dance language. *Annual Review of Entomology* 47: 917–949.

Eardley C D. 2004. Taxonomic revision of the African stingless bees (Apoidea: Apidae: Apinae: Meliponini). *African Plant Protection* 10(2): 63-96.

EFSA European Food Safety Authority. 2013c. *Guidance document on the risk assessment of plant protection products on bees (Apis mellifera, Bombus* spp. *and solitary bees)*. EFSA J. 11(NN), 268. DoI: 10.2903/j.efsa.2013.3295

Eickwort G C and Ginsberg H S. 1980. Foraging and mating behavior in Apoidea. *Annual Review of Entomology* 25: 421–46.

Eickwort G C and Eickwort K R. 1973. Notes on the nests of three wood-dwelling species of Augochlora from Costa Rica (Hymenoptera: Halictidae). *Journal of the Kansas Entomological Society* 46: 17-22.

Eltz T, Brühl C A, Imiyabir Z and Linsenmair K E. 2003. Nesting and nest trees of stingless bees (Apidae: Meliponini) in lowland dipterocarp forests of Sabah, Malaysia, with implications for forest management. *Forest Ecology and Management* 172: 301-313.

Engel M S and Dingemans-Bakels F. 1980. Nectar and pollen resources for stingless bees (Meliponinae, Hymenoptera) in Surinam (South America). *Apidologie* 11: 341–350.

Engels W and Imperatriz-Fonseca V L. 1990. Caste development, reproductive strategies, and control of fertility in honeybees and stingless bees. (*In*) Engels W (Ed.). Social insects: an evolutionary appraoch to castes and reproduction. Springer, Berlin Heidelberg, New York, pp 167–230.

Ercan N and Onus A N. 2003. The effects of bumble bees (*Bombus terrestris*

L.) on fruit quality and yield of pepper (*Capsicum annuum* L.) grown in an unheated green house. *Israel Journal of Plant Sciences*. 51: 275-283.

Erikson E H and Buchmann S L. 1983. *Electrostatics and pollination.* (In) Jones E E and Little R J (Eds). *Hand Book of Experimental Pollination Biology.* New York; Van Nostrand Reintold pp. 173-184.

Esch H, Esch I and Kerr W E. 1965. Sound: an element common to communication of stingless bees and to dances of the honeybee. *Science* 149: 320–321.

Esch H. 1967. Die Bedeutung der Lauterzeugung für die Verständigung der stachellosen Bienen, Z. Vergl. *Physiology* 56: 408–411.

Faegri K and Van der L P. 1966. *Principles of pollination ecology*. Pergamon Press.

Fain A and Flechtmann C H W. 1985. Mites of stingles bees *Melipona semingra*. *International Journal of Acarology* 11: 89-91.

Faustino C D, Silva M E V, Mateus S and Zucchi R. 2002. First record of emergency queen rearing in stingless bees (Hymenoptera, Apinae, Meliponini). *Insectes Sociaux* 49: 111–113.

Ferreira M G and Absy M L. 2013. Pollen analysis of the post-emergence residue of *Melipona (Melikerria) interrupta* Latreille (Hymenoptera: Apidae), created rationally in the Central Amazon. *Acta Botanica Brasilica* 27: 709-713.

Figueiredo M G, Bego L R and Nascimento F S. 2013. Foraging behavior of *Scaptotrigona depilis* (hymenoptera, apidae, meliponini) and its relationship with temporal and abiotic factors. *Sociobiology* 60(3): 277-282.

Fisher R M and Pomeroy N. 1989. Pollination of greenhouse muskmelons by bumble bees (Hymenoptera: Apidae). *Journal of Economic Entomology* 82(4): 1061-1066.

Flechtmann C H W, Phoon A C G and Camarcio C A. 1974. Mite associated with stingless bees from Brazil. (In) *Proceedings of the Fourth International Congress of Acarology,* 2-4 August, 1974, University of Austria, Sarlfelder. 315p.

Foitzik S and Heinze J. 1998. Nest site limitation and colony takeover in the ant *Leptothorax nylanderi*. *Behavioural Ecology* 9: 367–375.

Free J B and Butler C G. 1959. *Bumble bees*. Collins, London, 221 p.

Free J B and Ferguson A W. 1980. Foraging of bees on oil seed rape (*Brassica napus* L.) in relation to the stage of flowering of the crop and pest control. *Journal of Agricultural Sciences* 94: 151-154.

Free J B. 1993. *Insect pollination of crops*, 2nd edn. UK Academic Press, London, 544 p.

Fronk W D and Slater J A. 1956. Insect fauna of cucurbit flowers. *Journal of Kansas Entomological Society* 29: 141-146.

Gadhiya V C and Pastagia J J. 2019. Time spent by stingless bees, *Tetragonula laeviceps* for nectar and pollen collection from musk melon flower. *Journal of Entomology and Zoology Studies* 7(1): 498-500.

Gerardo P, Euan Q, Javier J G, Virginia R M, Javier I, Gustavo R V N and Manuei R. 2008. Comparative efficiency of *Nannotrigona perilampoides*, *Bombus impatiens* (Hymenoptera: Apoidea) and mechanical vibration on fruit production of enclosed habanero pepper. *Journal of Economic Entomology* 101(1): 132-138.

Giannini T C, Boff S, Cordeiro G D, Cartolano Jr. E A, Veiga A K, Imperatriz-Fonseca V L and Saraiva A M. 2015b. Crop pollinators in Brazil, a review of reported interactions. *Apidologie* 46: 209-223.

Gibbs J. 2009. Intergative taxonomy indentifies new (and old) species in the *Lasioglossum (Dialictus) tegulare* (Robertson) (Hymenoptera, Halictidae) species group. *Zootaxa* 2032: 1-38.

Gloag R, Heard T A, Beekman M and Oldroyd B P. 2008. Nest defence in a stingless bee: What causes fighting swarms in *Trigona carbonaria* (Hymenoptera, Meliponini)? *Insect Sociaux* 55: 387–391. https://DoI.org/10.1007/s00040-008-1018-1

Godfray H C J, Blacquiere T, Field L M, Hails R S, Petrokofsky G, Potts S G, Raine N E, Vanbergen A J and McLean A R. 2014. A restatement of the natural science evidence base concerning neonicotinoid insecticides and insect pollinators. *Proceedings of Royal Society of Britain* 281: 1-23. DoI: 10.1098/rspb.2014.0558

Gonzalez A R, Guererro A C and Arreola G S. 2016. *From extraction to meliponiculture-a case study of the management of stingless bees in the west central region of the Mexico*. In: Beekeeping and Bee Conservation- Advances in research. DoI: 10.5772/62654

Gonzalez V H and Griswold T. 2012. New species and previously unknown males of neotropical cleptobiotic stingless bees (Hymenoptera, Apidae, *Lestrimelitta*). *Caldasia* 34: 227–245.

Gonzalez-Acereto J A, Quezada-Euan J J G and Medina-Medina L A. 2006. New perspectives for stingless beekeeping in the Yucatan: results of an integral program to rescue and promote the activity. *Journal of Apicultural Research* 45(3): 234-239.

Goodale E, Beauchamp G, Magrath R D, Nieh J C and Ruxton G D. 2010. Interspecific information transfer influences animal community structure. *Trends in Ecology and Evolution* 25: 354-361. DoI: 10.1016/j.tree.2010.01.002

Goodell K and Thompson J D. 1997. Comparisons of pollen removal and deposition by honeybees and bumble bees visiting apple. *Acta Horticulturae* 437:103-107.

Goulson D, Chapman J W and Hughes W O H. 2001. Discrimination of unrewarding flowers by bees; direct detection of rewards and use of repellent scent marks. *Journal of Insect Behaviour* 14: 669–678.

Goulson D, Stout J C, Langley J and Hughes W O H. 2000. Identity and function of scent marks deposited by foraging bumble bees. *Journal of Chemical Ecology* 26: 2897–2911.

Goulson D. 2003. *Bumble bees, their behaviour and ecology*. Oxford University Press, Oxford.

Green J M. 1955. Bumble bees breed cotton too! *What's New in Crops and Soils* 7: 1-2

Greenleaf S S, Williams N M, Winfree R and Kremen C. 2007. Bee foraging ranges and their relationship to body size. *Oecologia* 153(3): 589-96.

Grueter C, von Zuben L, Segers F and Cunningham J. 2016. Warfare in stingless bees. *Insectes Sociaux* 63: 223–236. 10.1007/s00040-016-0468-0.

Gruter C and Leadbeater E. 2014. Insights from insects about adaptive social information use. *Trends in Ecology and Evolution* 29: 177-184. DoI: 10.1016/j.tree.2014.01.004

Guerrini A, Bruni R, Maietti S, Poli F, Rossi D and Paganetto G. Ecuadorian stingless bee (Meliponinae) honey: a chemical and functional profile of an ancient health product. *Food Chem*istry 114: 1413–1420.

Guibu L S, Ramalho M, Kleinert- Giovannini A and Imperatriz-Fonseca V L. 1988. Explora¸cao dos recursos florais por colonias de *Melipona quadrifasciata* (Apidae, Meliponinae). *Revista Brasilia de Biologica* 48: 299–305.

Gullan P J and Cranston P S. 2010. *The insects- an outline of Entomology.* 4th edition. Wiley-Blackwell 183: 308-336.

Gumbert G and Kunze J. 1999. Inflorescence height affects visitation behaviour of bees- a case study of aquatic plant community in Bolivia. *Biotropica* 31(3): 466-477.

Halcroft M, Spooner-Hart R and Neumann P. 2011. Behavioral defense strategies of the stingless bee, *Austroplebeia australis*, against the small hive beetle, *Aethina tumida. Insectes Sociaux* 58: 245-253. 10.1007/s00040-010-0142-x.

Halliwell B and Gutteridget J M C. 1984. Oxygen toxicity, oxygen radicals, transition metals and disease. *Journal of Biochemistry* 219:1–14.

Hamida. 1999. Enemies of bees. (In) Colin M E, Ball B V, Kilani M (eds). *Bee Disease Diagnosis,* 25th Ed. CIHEAM, Tunisia, pp 147-165.

Hart A G and Ratnieks F L W. 2002. Task-partitioned nectar transfer in stingless bees: work organisation in a phylogenetic context. *Ecological Entomology* 27: 163–168.

Hartfelder K and Engels W. 1989. The composition of larval food in stingless bees: evaluating nutritional balance by chemosystematic methods. *Insects Sociaux* 36: 1–14.

Heard T A and Dollin A. 2000. Stingless beekeeping in Australia, snapshot of an infant industry. *Bee World* 82: 116-125.

Heard T A, Vithanage V and Chacko E K. 1990. Pollination biology of cashew in the Northern Territory of Australia. *Australian Journal of Agricultural Research* 41:1101–1114.

Heard T A. 1999. The role of stingless bees in crop pollination. *Annual Review of Entomology* 44: 183-206.

Heard T A. 1988. Propagation of hives of *Trigona carbonaria* Smith (Hymenoptera: Apidae). *Australian Journal of Entomological Society* 27: 303-304.

Heard T A and Hendrikz J K. 1993. Factors influencing flight activity of colonies of the stingless bee *Trigona carbonaria* (Hymenoptera: Apidae). *Austalian Journal Zoology* 41: 343–53.

Hebert P D N, Penton E H, Burns J M, Janzen D H and Hallwachs W. 2004. Ten species in one: DNA barcoding reveals cryptic species in the neotropical butterfly *Astraptes fulgerator*. *Proceedings of Natural Academic Sciences USA* 101: 14813-14817.

Hebert P D N, Ratnasingham S and de Waard J R. 2003. Barcoding animal life: cytochrome *c* oxidase subunit 1 divergences among closely related species. *Proceedings of Royal Society of London Biological Sciences* 270(1): 96-99.

Hebling N J, Kerr W E and Kerr F S. 1964. Divisão de trabalho entre operárias de *Trigona (Scaptotrigona) xanthotricha* Moure. *Papéis Avuls Zool., São Paulo* 16(13):115-127.

Heemert C V, Ruitjer A D, Eijnde J V and Steen J V. 1986. Year-round production of bumble bee colonies for crop pollination. *Bee world* 71(2): 54-56.

Heithaus E R. 1979a. Flower-feeding specialization in wild bee and wasp communities in seasonal neotropical habitats. *Oecologia* 42: 179-194.

Heithaus E R. 1979 b. Flower visitation records and resource overlap of bees and wasps in northwest Costa Rica. *Brenesia* 16: 9-52.

Hempel P S. 1998. Parasites in social insects. Princeton University Press, Princeton. 34p.

Hendrichs P R. 1941. The cultivation of indigenous bees in the State of Guerrero. *Mexico Antiguo* 5: 365–73.

Herbert S F. 1945. The wax of stingless bees and the use to which it has been put. *Journal of the New York Entomological Society* 35: 137-144.

Herbst S T. 2001. The New Food Lover's Companion: Comprehensive Definitions of Nearly 6,000 Food, Drink, and Culinary Terms. Barron's Cooking Guide. Hauppauge, NY: Barron's Educational Series.

Higo H A, Rice N D, Winston M L and Lewis B. 2004. Honeybee distribution and potential for supplementary pollination in commercial tomato greenhouses during winter. *Journal of Economic Entomology* 97: 163-170.

Hilario S D and Imperatriz F. 1999. Flight activity and colony strength in stingless bee *Melipona bicolor bicolour* (Apidae, Meliponinae). *Revista Brasileira de Biologica* 60(2): 299-306.

Hilario S D, Ribeiro M F and Imperatriz-Fonseca V L. 2012. Can climate shape flight activity patterns of *Plebeia remota* (Hymenoptera, Apidae)? Iheringia. Série *Zoologia* 102: 269-276.

Hines H M, Cameroon S A and Deans A R. 2007. Nest architecture and foraging behaviour in *B. pullatus* (Hymenoptera: Apidae), with comparisons to other tropical bumble bees. *Journal of the Kansas Entomological Society* 80(1): 1-15.

Hobbs G A, Nummi W O and Virostek J F. 1961. Food gathering behavior of honeybees, bumble bees and leaf cutting bees (Hymenoptera: Apoidea) in Alberta. *Canadian Entomology* 93: 409-419.

Hogendoorn K, Steen Z and Schwarz M P. Native Australian carpenter bees as a potential alternative to introducing bumble bees for tomato pollination in greenhouses. *Journal of Apicultural Research* 39(1-2): 67–74.

Hogendoorn K. 2004. On promoting solitary bee species for use as crop pollinators in greenhouses. *Solitary Bees: Conservation, Rearing and Management for Pollination*. B Freitas M, and Pereira J O P (Eds). pp 213–221. Imprensa Universitária, Fortaleza, Brazil.

Holldobler B and Wilson E O. 1978. The multiple recruitment systems of the African weaver ant, *Oecophylla longinoda* (Latreille) (Hymenoptera: Formicidae). *Behavioural Ecology and Sociobiology* 3: 19–60.

Holldobler B and Wilson E O. 1990. *The ants*. The Belknap Press of Harvard University Press, Cambridge, Massachusetts.

Holm S N. 1966. Problems of the domestication of bumble bees. *Bee World* 47: 179-186.

Hossam F and Shara A. 2015. The origin of honeybee life- a viewpoint. *Journal of Entomology and Zoology Studies* 3(1): 239-241.

Hrncir M, Jarau S, Zucchi R and Barth F G. 2002. *Dependence of sound signals in Melipona seminigra with factors other than food location*. (In) Billen J (Ed.). XIV Int. Congr. IUSSI: The golden jubilee proceedings, Science Council of Japan, Entomological Society of Japan, Hokkaido University Coop, Hokkaido University, Sapporo, Japan, p 124.

Hrncir M, Jarau S, Zucchi R and Barth F G. 2004. On the origin and properties of scent marks deposited at the food source by a stingless bee, *Melipona seminigra* Friese 1903. *Apidologie* 35: 3–13.

Hrncir M, Jarau S, Zucchi R and Barth F G. 2000. Recruitment behaviour in stingless bees, *Melipona scutellaris* and *M. quadrifasciata*. II. Possible mechanisms of communication. *Apidologie* 31: 93–113.

http://entnemdept.ufl.edu/creatures/misc/bees/halictid_bees.htm

http://www.fao.org/3/i0842e/i0842e07.pdf

Hubbell S P and Johnson L K. 1977. Competition and nest spacing in a tropical stingless bee community. *Ecology* 58: 949-963.

Hurd P D. 1957. Notes on the autumnal emergence of the vernal desert bee, *Hesperapis fulvipes* Crawford (Hymenoptera, Apoidea). *Journal of Kansas Entomological Society* 30: 10-15.

Imperatriz-Fonseca V L. 1973. Miscellaneous observations on the behaviour of *Schwarziana quadripunctata*. *Boletim de Zoologia e Biologia Marinha. Sao Paulo* 30: 633–640.

Imperatriz-Fonseca V L and Zucchi R. 1995. Virgin queens in stingless bees (Apidae, Meliponinae) colonies: a review. *Apidologie* 26: 231–244.

Imperatriz-Fonseca V L, Criz-Landim C and Silva de Moraes R L M. 1997. Dwarf queens in *Nannotrigona testaceicornis* (Apidae, Meliponinae, Trigonini): behaviour, exocrine gland morphology and reproductive status. *Apidologie* 28: 113–122.

Inoue T, Nakamura K, Salmah S and Abbas I. 1993. Population-dynamics of animals in unpredictably- changing tropical environments. *Journal of Biosciences* 18: 425–455.

Inoue T, Sakagami S F, Salmah S and Nukmal N. 1984. Discovery of successful absconding in the stingless bee *Trigona (Tetragonula) laeviceps*. *Journal of Apicultural Research* 23: 136–142.

Inoue T, Sakagami S F, Salmah S and Yamane S. 1984. The process of colony multiplication in the Sumatran stingless bee *Trigona (Tetragonula) laeviceps*. *Biotropica* 1: 100–111.

Inoue T, Salmah S, Abbas I and Yusuf E. 1985. Foraging behavior of individual workers and foraging dynamics of colonies of three Sumatran stingless bees. *Researches on Population Ecology* 27: 373-392.

Ish-Am G and Eisikowitch D. 1998. Mobility of honeybees (Apidae, *Apis mellifera* L.) during foraging in Avocado orchards. *Apidologie* 29(3): 209-219.

Iwama S and Melhem T S. 1979. The pollen spectrum of the honey of *Tetragonisca angustula angustula* Latreille (Apidae, Meliponinae). *Apidologie* 10(3): 275-295.

Iyengar E V. 2008. Kleptoparasitic interactions throughout the animal kingdom and a re-evaluation, based on participant mobility, of the conditions promoting the evolution of kleptoparasitism. *Biological Journal of Linneaus Society* 93: 745–762.

Jacob C R O, Soares H M, Carvalho S M, Nocelli R C F and Malaspina O. 2013. Acute toxicity of fipronil to the stingless bee *Scaptotrigona postica* latreille. *Bullettin of Environment Contamination and Toxicology* 90(1): 69-72. DoI: 10.1007/s00128-012-0892-4

Jaffe R, Pope N, Carvalho A T, Maia U M, Blochtein B, Carvalho C A L, Carvalho-Silze G A, Freitas B M, Menezes C, Ribeiro M F, Venturieri G C and Imperatriz- Fonseca V L. 2015. Bees for development: Brazilian survey reveals how to optimize stingless beekeeping. *PLOS ONE* 10(3). DoI: 10.1371/journal.pone.0121157

Jalil M A A, Kashmuri A R and Hadi H. 2017. Stingless bee honey, the natural wound healer: a review. *Skin Pharmacolology Physiology* 30: 66–75.

Jarau S, Hrncir M, Ayasse M, Schulz C, Francke W, Zucchi R and Barth F G. 2004. A stingless bee (*Melipona seminigra*) marks food sources with a pheromone from its claw retractor tendons. *Journal of Chemical Ecology* 30: 793-804.

Jarau S, Hrncir M, Zucchi R and Barth F G. 2002. Foot print pheromones used to mark food sources by stingless bees. (In) Billen J (Ed). XIV Int. Congr. IUSSI: the golden jubilee proceedings, Hokkaido University Coop, Hokkaido University, Sapporo, Japan, p 16.

Jarau S, Morawetz L, Reichle C, Gruber M H, Huber W and Weissenhofer A. 2009. Corbiculate bees of the Golfo Dulce Region, Costa Rica. Wien, Verein zur Förderung der Tropenstation La Gamba, Costa Rica.

Jayalekshmi C R. 2015. *Pests and diseases of stingless bee Tetragonula iridipennis Smith (Meliponinae: Apidae)*. PhD Thesis, Department of Agricultural Entomology, College of Agriculture, Vellayani, Thiruvananthapuram, Kerala.

Jobiraj T and Narendran T C. 2004. Revised key to the world species of *Lisotrigona moure* (Hymenoptera: Apoidea: Apidae) with description of a new species from India. *Entomon* 29: 39-43.

Johnson L K. 1987. The pyrrhic victory of nest-robbing bees: did they use the wrong pheromone? *Biotropica* 19:188–189.

Johnson R M .2015. Honeybee toxicology. *Annual Review of Entomology* 60: 415–434.

Johnson R M, Ellis M D, Mullin C A and Frazier M. 2010. Pesticides and honeybee toxicity – USA. *Apidologie* 41(3): 312–331.

Juliani L. 1967. A descricao do ninho e alguns dados biológicos sobre a abelha *Plebeia julianii* Moure, 1962 (Hymenoptera: Apoidea). *Revista Brasilia de Entomologica* 12: 31–58.

Kaesar T. 2010. Large carpenter bees as agricultural pollinators. *Psyche- A Journal of Entomology* DoI:10.1155/2010/927463

Kakutani T, Inoue T, Tezuka T and Maeta Y. 1993. Pollination of strawberry by the stingless bee, *Trigona minangkabau*, and the honeybee, *Apis mellifera*: an experimental study of fertilization efficiency. *Researches on Population Ecology* 35: 95–111.

Karuppasamy V K and Jeyaraaj R. 2016. Floral sources for stingless bees (*Tetragonula iridipennis*) in Nellithurai Village, Tamil Nadu, India. *Ambient Science* 3(2): 17-21. DoI:10.21276/ambi.2016.03.2.ra04

Kashyap L. 2008. *Domicilation of bumble bees (Bombus sp.) and to study resource partitioning with honeybees*. MSc thesis, Dr Y S Parmar University of Horticulture and Forestry, Nauni, Solan, Himachal Pradesh.

Katayama E. 1987. Utilization of honeybees as pollinators for strawberries in plastic greenhouses. *Honeybee Science* 8: 147-150.

Kazuhiro A. 2004. Attempts to introduce stingless bees for the pollination of crops under greenhouse conditions in Japan. National Institute of Livestock and Grassland Science, Laboratory of Apiculture, Tsukuba, Ibaraki 305–0901, Japan. pp. 1-9.

Keasar T, Sadeh A, Shilo M and Ziv Y. 2007. Social organization and pollination efficiency in the carpenter bee *Xylocopa pubescens* (Hymenoptera: Apidae: Anthophorinae). *Entomologia Generalis* 29(2–4): 225–236.

Kelly N, Farisya M S N, Kumara T K and Marcela P. 2014. Species diversity and external nest characteristics of stingless bees in Meliponiculture. *Pertanika Journal Tropical Agricultural Science* 37(3): 293-298.

Kendal R L, Coolen I, van Bergen Y and Lal K N. 2005. Trade-offs in the adaptive use of social and asocial learning. *Advances Study Behaviour* 35: 333-379. DoI: 10.1016/ s0065-3454(05)35008-x

Kent R B. 1984. Mesoamerican stingless beekeeping. *Journal of Cultural Geography* 4: 14-28. DoI: 10.1080/0887363840 9478571

Kerr W E, Carvalho G A and Nascimento V A. 1996. Abelha Uruçu: Biologia, Manejo e Conservação – Belo Horizonte-MG: Acangaú: il., (Coleção Manejo da vida silvestre; 2).

Kerr W E and Cruz C D. 1961. Funções diferentes tomadas pela glândula mandibular na evolução das abelhas e em *Trigona* (*Oxytrigona*) *tataira* em especial. *Revista Brasileira de Biologica* 21: 1–16.

Kerr W E and Esch H. 1965. Comunicação entre as abelhas sociais brasileiras e sua contribuicao para o entendimento da sua evolução. *Ciencia Cultura* 17: 529–538.

Kerr W E and Rocha R. 1988. Comunicação em Melipona rufiventris e Melipona compressipes. *Ciencia Cultura* 40: 1200–1203.

Kerr W E, Blum M and Fales H M. 1981. Communication of food sources between workers of *Trigona (Trigona) spinipes*. *Revista Brasileira de Biologica* 41: 619–623.

Kerr W E, Ferreira A and Simoes de Mattos N. 1963. Communication among stingless bees additional data (Hymenoptera: Apidae). *Journal of New York Entomological Society* 71: 80–90.

Kerr W E. 1948. Estudos sôbre o gênero Melipona. Doctoral thesis, Universidade de São Paulo.

Kerr W E. 1960. Evolution of communication in bees and its role in speciation. *Evolution* 14: 326– 327.

Kerr W E. 1969. Some aspects of the evolution of social bees. *Evolution Biology* 3: 119–175.

Kerr W E. 1972. Orientação pelo sol em *Trigona spinipes* Ciencia. *Cult Supplement* 4: 341–342.

Kerr W E. 1973. Sun compass orientation in the stingless bees, *Trigona* (*Trigona*) *spinipes* (Fabricius, 1793) (Apidae). *An Acadamecia Ciencia de Brasileira* 45: 301–308.

Kerr W E. 1994. Communication among *Melipona* workers (Hymenoptera: Apidae). *Journal of Insect Behaviour* 7: 123–128.

Kerr W E. 1990. Why are workers in social Hymenoptera not males? *Revista Brasilia de Genetics*. 13: 133–136.

Kerr W E. 1997. Sex determination in honeybees (Apinae and Meliponinae) and its consequences. *Brazilian Journal of Genetics* 20: 601–611.

Kiatoko N, Kumar R S and Langevelde F. 2016. A vertical compartmented hive design for reducing post-harvest colony losses in three afro tropical stingless bee species (Apidae: Meliponinae). *International Journal of Development*

Research 6(8): 9026-9034.

Kiatoko N, Raina S K S, Muli E, Mueke J, Nkoba K, Raina S K S and Mueke J. 2014. Enhancement of fruit quality in *Capsicum annum* through pollination by *Hypotrigona griboDoI* in Kakamega, Western Kenya. *Entomological Science* 17(28): 106-110.

Kiatoko N. 2012. *Distribution, behavioural biology, rearing and pollination efficiency of five stingless bee species (Apidae: Meliponinae) in Kakamega forest, Kenya*. Ph D thesis, Pure and Applied Science of Kenyatte University, Kenya, 140-226 pp.

King M J and Buchmann S L. 1995. Bumble bee initiated vibration release mechanism of *Rhododendron* pollen. *American Journal of Botany* 82(11): 1407-1411.

Kirchner W H. 1994. Hearing in honeybees: the mechanical response of the bee's antenna to near field sound. *Journal of Comparative Physiology A* 175: 261–265.

Klein A M, Steffan-Dewenter I and Tscharntke T. 2003. Fruit set of highland coffee increases with the diversity of pollinating bees. *Proceedings of Royal Society of London B* 270: 955–961.

Kleinert-Giovannini A and Imperatriz- Fonseca V L. 1987. Aspects of the trophic niche of *Melipona marginata marginata* Lepeletier (Apidae, Meliponinae). *Apidologie* 18: 69–100.

Klumpp J. 2007. *Australian stingless bees: a guide to sugarbag beekeeping*. Brisbane, Earthling Enterprises Pty Ltd, 110 p.

Koch H. 2010. Combining morphology and DNA barcoding resolves the taxonomy of Western Malagasy *Liotrigona* Moure, 1961 (Hymenoptera: Apidae: Meliponini). *African Invertebrates* 51(2): 413-421.

Kshirsagar K K and Chauvan R M. 1977. Preliminary observations on oviposition behavior in *Trigona (Tetragona) iridipennis* Smith. *Indian Bee Journal* 39 (1-4): 22-23.

Kumar M S, Singh A J A R and Ahugunuth G. 2012. Traditional Beekeeping of stingless bee (*Trigona* sp.) by Kani Tribes of Western Ghats, Tamil Nadu, India. *Indian Journal of Traditional Knowledge* 11(2): 342-345.

Kumar V K. 2014. Nest and colony characters of *Trigona (Lepidotrigona) ventralis* var. *arcifera* Cockerell from North East India. *Asian Journal of Conservation Biology* 3(1): 90-93.

Kustiawan P M, Songchan P, Enos T A and Chanpen C. 2014. *In vitro* cytotoxicity of Indonesian stingless bee products against human cancer cell lines. *Asian Pacific Journal of Tropical Biomedicine* 4(7): 549-556.

Kwapong P, Aidoo K, Combey R and Karikari A. 2010. *Stingless bees: Importance, Management and Utilization*. A training manual for stingless beekeeping. Unimax Macmillan Ltd, Ghana: Accra North, 72 p.

Kwon Y J and Saeed S. 2003. Effect of temperature on the foraging activity of *Bombus terrestris* L. (Hymenoptera: Apidae) on greenhouse hot pepper (*Capsicum annum*). *Applied Entomology and Zoology* 38: 275-280.

Lacerda L M and ZucchI R. 1999. Behavioral alterations and related aspects in queenless colonies of *Geotrigona mombuca* (Hymenoptera, Apidae, Meliponinae). *Sociobiology* 33(3): 277-288.

Larkin L, Neff J L and Simpson B B. 2008. The evolution of specificity: Pollen host choice and diet breadth of Andrena bees. *Apidologie* 39: 771-799.

Lourenco C T, Carvalho S M, Malaspina O and Nocelli R C F. 2012. Oral toxicity of fipronil insecticide against the stingless bee *Melipona scutellaris* (Latreille, 1811). *Bullettin of Environment Contamination and Toxicology* 89(4): 921-924.

Leao K, Queiroz A, Veiga J, Contrera F and Venturieri G. 2017. Colony development and management of the stingless bee *Scaptotrigona aff. postca* (Apidae: Meliponini) using different hive models. Sociobiology 63(4): 1038.

Light J. 1994. Bumble bee pollination. *Insecta* 7: 45-46.

Lima M A P, Pires C S P, Guedes R N C and Campos L A O. 2013. Lack of lethal and sublethal effects of Cry1Ac Bt-toxin on larvae of the stingless bee *Trigona spinipes*. *Apidologie* 44: 21–28.

Lindauer M and Kerr W E. 1960. Communication between the workers of stingless bees. *Bee World* 41: 65-71.

Lindauer M and Kerr W E. 1958. Die gegenseitige Verständigung bei den stachellosen Bienen, Z. Vergl. *Physiology* 41: 405–434.

Lindeman M and Araya A. 2012. New stingless bee boxes and shallow frames. http://costaricaapiaries.com/blog/?p=73

Lindstrom A, Korpela S and Fries I. 2008. Horizontal transmission of *Paenibacillus larvae* spores between honeybee (*Apis mellifera*) colonies through robbing. *Apidologie* 39: 15–522.

Linsley E G. 1958. The ecology of solitary bees. *Hilgardia* 27: 543–599.

Lobreau-Callen D, Thomas Ale, Darchen B and Darchen R. 1990. Quelques facteurs d´eterminant le comportement de butinage d'Hypotrigona pothieri (Trigonini) dans la vegetation de Coted Ivoire. *Apidologie* 21: 69–83.

Luz C F P and Barth O M. 2001. *Melissopalynological observations in a mangrove area next to Rio de Janeiro, Brazil*. (*In*) Goodman D K, Clarke R T (Eds). Proceedings of the IX Palynological Congress. American Association of Stratigraphic Palynologists Foundation, Houston, pp 489–492.

Luz C F P D, Fernandes-Salamao T M, Lage L G A, Resende H C, Tavares M G and Campos L A D O. 2011. Pollen sources for *Melipona capixaba* Moure & Camargo- an endangered Brazilian stingless bee. *Psyche- a journal of Entomology* p 7. https://DoI.org/10.1155/2011/107303

Macfarlane R P, Griffin R P and Read P E C. 1983. Bumble bee management options to improve 'Grassland Pawera' red clover seed yields. *Proceedings of New Zealand Grassland Association* 44: 47–53.

Macfarlane R P, Patten K D, Mayer D F and Shanks C H. 1994b. Evaluation of commercial bumble bee colonies for cranberry pollination. *Melanderai* 50: 13-19.

MacKenzie K E. 1994. The foraging behaviour of honeybees (*Apis mellifera* L.) and bumble bees (*Bombus* spp.) on cranberry (*Vaccinium macrocarpon* Ait). *Apidologie* 25: 375-383.

Mackenzie K. 2009. Pollination practices and the use of bees in the vaccinum crops. *Acta Horticulturae* 810: 207-213.

Maeta Y, Tezuka T, Nadano H and Suzuki K. 1992. Utilization of the Brazilian stingless bee, *Nannotrigona testaceicornis* as a pollinator of strawberries. *Honeybee Science* 13: 71-78.

Malagodi-Braga K S and Kleinert A M P. 2004. Could *Tetragonisca angustula* Latreille (Apinae, Meliponini) be used as strawberry pollinator in greenhouses?

Australian Journal of Agricultural Research 55: 771–773.

Malkowski S R, Faraj B H and Schwartz-Filho D L. 2006. *Eficiência de garrafas-iscas na captura de enxames de Tetragonisca angustula (Latreille, 1811) (Hymenoptera, Apidae), Anais do 16º*. Congresso Brasileiro de Apicultura e 2º Congresso Brasileiro de Meliponicultura Aracaju, SE.

Massaro F C, Brooks P R, Wallace H M and Russell F D. 2011. Cerumen of Australian stingless bees (*Tetragonula carbonaria*): gas chromatography-mass spectrometry fingerprints and potential anti-inflammatory properties. *Naturwissenschaften* 98: 329-337.

Mattu V K, Mattu N, Verma L R and Lakhanpal T N. 1988. *Pollen spectrum of honey from Apis cerana colonies in Himachal Pradesh India*. Proceedings of the Fourth International Conference on Apiculture in Tropical Climates, Cairo, Egypt, pp 146-153.

Mattu V K. 1989. *Scope and strategies for Apicultural development in Himachal Pradesh*. (*In*) Honeybees in mountain agriculture. Verma L R (Ed.). New Delhi, Oxford & IBH Publ. Co. Pvt Ltd, New Delhi.

Mayer D F, Johansen C A and Lude J D. 1987. *Pear pollination in Washington*. St. Univ. Crop Extension Proceedings on Pear production in the Pacific Northwest, 266 p.

Mayer D F, Johansen C A and Baird C R. 1999. *How to reduce bee poisoning from Pesticides*. Washington State University.

McGregor S E. 1976. Insect pollinators on cultivated crop plants. *Agricultural Hand book*, p 496.

McGregor S E. 1976. Insect pollination of cultivated crop plant. *Agricultural Handbook*. USDA-ARS, Washington D C, New York, p 496.

Mcguire C M. 1999. *Passiflora incarnata* (Passifloraceae): a new fruit crop. *Economic Botany* 53(2): 161–176.

Medeiros V de F L P, Azevedo Í M, Rêgo A C M, Egito E S T do, Araújo-Filho I and Medeiros A C. 2016. Antibacterial properties and healing effects of *Melipona scutellaris* honey in MRSA-infected wounds of rats. *Acta Cirurgica Brasileira* 31(5): 327–332.

Medler J T. 1956. Principles and methods for the utilization of bumble bees in cross-pollination of crops. *Proceedings of 10th International Congress on Entomology* 4: 973-981.

Meisels and Chiasson. 1997. Effectiveness of *B. impatiens* as pollinators of greenhouse sweet peppers (*Capsicum annuum* L.). *Acta Horticulturae* 437: 425-429.

Menezes C, Coletto-Silva A, Gazeta G S and Kerr W E. 2009. Infestation by *Pyemotes tritici* (Acari, Pyemotidae) causes death of stingless bee colonies (Hymenoptera: Meliponina). *Genetics and Molecular Research* 8(2): 630-634.

Menezes C, Vollet-Neto A and Fonseca V. 2013. An advance in the *in vitro* rearing of stingless bee queens. *Apidologie* 44(5): 491-500.

Menezes C, Vollet-Neto A, Contrera F A F L, Venturieri G C and Imperatriz-Fonseca V L. 2013. The role of useful microorganisms to stingless bees and stingless beekeeping. (*In*) *Pot-Honey: A Legacy of Stingless Bees*. New York, Springer, pp 153-171.

Michelsen A, Kirchner W H and Lindauer M. 1986b. Sound and vibrational signals in the dance language of the honeybee, *Apis mellifera*. *Behavioural*

Ecology and Sociobiology 18: 207–212.

Michelsen A. 2003. Signals and flexibility in the dance communication of honeybees. *Journal of Comparative Physiology A*. 189: 165–174.

Michener C D. 1946. Notes on the habits of some Panamanian stingless bees (Hymenoptera, Apidae). *Journal of New York Entomological Society* 54: 179–197.

Michener C D. 2000. *The bees of the world*. The Johns Hopkins University Press, Baltimore.

Michener C D. 2013. *The Meliponini*. (*In*) Vit P, Pedro S R M and Roubik DW (Eds). *Pot-Honey: A legacy of stingless bees.* Springer, New York, pp 3-17.

Michener C D. 1974. *The social behavior of the bees - a comparative study*. Cambridge, Massachusetts, Belknap Press of Harvard University Press, 404 p.

Michener C D. 2007. *The bees of the world. Second edition*. Johns Hopkins University Press, Baltimore.

Michez D, Patiny S, Rasmont P, Timmermann K and Nicolas J Vereecken. 2008. Phylogeny and host-plant evolution in Melittidae s.l. (Hymenoptera: Apoidea). *Apidologie* 39(1): 146-162.

Milborrow B V, Kennedy J M and Dollin A. 1987. Composition of wax made by the Australian stingless bee *Trigona australis*. *Australian Journal of Biological Sciences* 40: 15-26.

Minckley R L, Cane J H and Kervin L. 2000. Origins and ecological consequences of pollen specialization among desert bees. *Proceedings of Royal Society of London B*. 267: 265–271.

Moraes S S, Bautista A R L and Viana B F. 2000. Avaliacao da toxicidade aguda (DL50 e CL50) de inseticidas para *Scaptotrigona tubiba* (Smith) (Hymenoptera: Apidae): via de contato. *Annals of Entomology Society of Brasilia* 29(1): 31-37.

Morgado L N, Andrade R C, Lorenzon M C A and Goncalves-Esteves V. 2011. Pollen used by *Tetragonisca angustula* Latreille (Apidae: Meliponina). *Acta Botanica Brasilica* 25(4): 932-934.

Morgner F. 1977. Rearing bumble bees and leaf cutter bees, and their use for pollination in small propagating units. *Journal of Economic Entomology* 71(2): 169-173.

Morse R A. 1976. Efficient use of honeybees in the orchard. *Proceeding of the Annual Meeting New York Horticultural Society* 1021: 61-64.

Moure J S. 1961. A preliminary supra-specific classification of the Old World Meliponinae bees (Hymenoptera-Apoidea). *Studia Entomology Rio de Jenerio* 4: 181-242.

Muller A. 1996. Host-plant specialization in western palearctic anthidiine bees (Hymenoptera: Apoidea: Megachilidae). *Ecological Monograph* 66: 235–257.

Muller A and Kuhlmann M. 2008. Pollen hosts of western palaearctic bees of the genus *Colletes* (Hymenoptera: Colletidae) – the Asteraceae paradox. *Journal of Biological Linnaeaus Society* 95: 719–733.

Muller F. 1874. The habits of various insects. *Nature* 10:102–103.

Mullin C A, Chen J, Fine J D, Frazier M T and Frazier J L. 2015. The formulation makes the honeybee poison. *Pesticide Biochemical Physiology* 120: 27–35.

Muthuraman M and Saravanan P A. 2004. Utilization of stingless bees for crop pollination. *Indian Bee Journal* 66: 58-64.

Nagaland Beekeeping Honey Mission. 2018. *Proceedings of Honeybee day 5th May 2018*.

Namwong A. 2003. *The Effectiveness of Trigona laeviceps Smith (Hymenoptera : Apidae) for increasing yield of dragon fruits (Hylocereus spp.)*. M Sc thesis.

Nates-Parra G. 2005. Abejas silvestres y polinización. Manejo Integr. *Plagas Agroecology*. 95: 7-20.

Nayak P P and Prakash J. 2017. Molecular characterization of the indigenous stingless bees (*Tetragonula* spp. Complex) using ISSR Marker from Southern Peninsular India. *Neotropical Entomology* 47: 106-117.

Nayak R. 2018. *Studies on bumble bee pollination in kiwifruit (Actinidia deliciosa* Chev.). M Sc thesis, Dr Y S Parmar University of Horticulture and Forestry, Nauni, Solan.

Ndungu N N, Kiatoko N, Ciosi M, Salifu D, Nyansera D, Masiga D and Raina S K. 2017. Identification of stingless bees (Hymenoptera: Apidae) in Kenya using Morphometrics and DNA barcoding. *Journal of Apicultural Research* 56(4): 341-353.

Neiswinder R B. 1956. Pollination of glass house tomatoes by honeybees. *Bee Craft* 72(8): 228-231.

Nicodemo D, Malheiros E B, Jong D D and Couto R H N. 2013. Enhanced production of parthenocarpic cucumbers pollinated with stingless bees and Africanized honeybees in greenhouses. *Semina: CiênciasAgrárias Londrina* 34(6): 3625-3634.

Nieh J C and Roubik D W. 1995. A stingless bee (*Melipona panamica)* indicates food location without using a scent trail. *Behaviour Ecology and Sociobiology* 37: 63–70.

Nieh J C and Roubik D W. 1998. Potential mechanisms for the communication of height and distance by a stingless bee, *Melipona panamica*. *Behavioural Ecology and Sociobiology* 43: 387–399.

Nieh J C and Tautz J. 2000. Behaviour-locked signal analysis reveals weak 200–300 Hz comb vibrations during the honeybee waggle dance. *Journal of Experimental Biology* 203: 1573–1579.

Nieh J C, Contrera F A L and Nogueira-Neto P. 2003a. Pulsed mass-recruitment by a stingless bee, *Trigona hyalinata*. *Proceedings of Royal Society of London B* 270: 2191–2196.

Nieh J C, Contrera F A L, Rangel J and Imperatriz- Fonseca V L. 2003c. Effect of food location and quality on recruitment sounds and success in two stingless bees, *Melipona mandacaia* and *Melipona bicolour*. *Behavioural Ecology and Sociobiology* 55: 87–94.

Nieh J C, Ramirez S and Nogueira-Neto P. 2003. Multi-source odor-marking of food by a stingless bee, *Melipona mandacaia*. *Behavioural Ecology and Sociobiology* 54: 578-586.

Nieh J C. 1998. The food recruitment dance of the stingless bee, *Melipona panamica*. *Behavioural Ecology and Sociobiology* 43: 133–145.

Nieh J C. 1998b. The role of a scent beacon in the communication of food location in the stingless bee, *Melipona panamica*. *Behavioural Ecology and Sociobiology* 43: 47–58.

Nieh J C. 1999. Stingless-bee communication. *American Science* 87: 428–435.

Nieh J C. 2004. Recruitment communication in stingless bees (Hymenoptera,

Apidae, Meliponini). *Apidologie* 35: 159-182.

Nisha M M. 2002. *Management of stingless bee Trigona iridipennis Smith (Meliponinae: Apidae) in the homesteads of Kerala*. MSc (Ag.) thesis, Kerala Agricultural University, Thrissur, pp 62-63.

Nishio E K, Ribeiro J M, Oliveira A G, Andrade C G T J, Proni E A, Kobayashi R K T and Nakazato G. 2016. Antibacterial synergic effect of honey from two stingless bees: *Scaptotrigona bipunctata* Lepeletier, 1836, and *S. postica* Latreille, 1807. *Scientific Reports* 6(1): 21641.

Nkoba K, Raina S K, Muli E, Mithöfer K and Mueke J. 2012. Species richness and nest dispersion of some tropical meliponine bees (Apidae: *Meliponinae*) in six habitat types in Kakamega forest, western Kenya. *International Journal of Tropical Insect Science* 32: 194-202.

Nobre da Cruz C B, Pieri F A, Carvalho-Zilse G A, Orlandi P P, Nunes-Silva C G and Leomil L. 2014. Antimicrobial activity of honeys from two stingless honeybee species and *Apis mellifera* (Hymenoptera: Apidae) against pathogenic microorganisms. *Acta Amazonica* 44(2): 287–290.

Nogueira-Neto P. 1954. Notas bionômicas sobre Meliponíneos III- Sobre a enxameagem (Hymenoptera- Apoidea). *Archivos do Museu Nacional* 42: 419–452

Nogueira-Neto P. 1970. *Behavior problems related to the pillages made by some parasitic stingless bees (Meliponinae, Apidae)*. In: Aronson L R (Ed.). Development and evolution of behavior: essays in memory of TC Schneirla. W. H. Freeman, San Francisco, pp 416–434.

Nogueira-Neto P. 1997. *Vida e Criação de Abelha sem Ferrao*. Editora Nogueirapis, Sao Paulo, p 447.

Nunes – Silva, Hrncir M, Inesda Silva C, Roldao Y S and Imperatriz – Fonseca V L. 2013. Stingless bees, *Melipona fasciculata*, as efficient pollinators of egg plant (*Solanum melongena*) in greenhouse. *Apidologie* 44: 537-546.

Nunes T M, Heard T A, Venturier G C and Oldroyd B P. 2014. Emergency queens in *Tetragonula carbonaria* (Smith, 1854) (Hymenoptera: Apidae: Meliponini). *Austral Entomology* 54(2): 154-158.

Occhiuzzi P. 2000. Stingless bee pollinate greenhouse capsicum. *Aussie bee* 13:15.

Oliveira R C, Menezes C, Ademilson E E S and Fonseca V V I. 2012. Trap-nests for stingless bees (Hymenoptera, Meliponini). *Apidologie* 44(1): 29-37.

Ollerton J, Killick A, Lamborn E, Watts S and Whiston M. 2007. Multiple meanings and modes: on the many ways to a generalist flower. *Taxon* 56: 717–728.

Ozbalci B, Boyaci I H, Topcu A, Kadilar C and Tamer U. 2013. Rapid analysis of sugars in honey by processing Raman spectrum using chemometric methods and artificial neural networks. *Food Chemistry* 136(3–4): 1444–1452.

Pallavi K. 2016. *Influence of hive design on seasonal performance of stingless honeybee, Tetragonula irridipennis (Smith)*. MSc (Agri) Thesis, University of Agricultural Sciences, Bengaluru, India.

Palma G, Quezada-Euán J J G, Reyes-Oregel V, Meléndez V and Moo-Valle H. 2008. Production of greenhouse tomatoes (*Lycopersicon esculentum*) using *Nannotrigona perilampoides*, *Bombus impatiens* and mechanical vibration (Hym : Apoidea). *Journal of Applied Entomology* 132: 79-85.

Parker F D and Torchio P P. 2007. *Bombus terrestris* cutting a flower to rob its nectar. *Beekeeping Information Index* 17: 13-15.

Partap T. 1992. *Honey plant sources in mountain areas- honeybee in mountain in Agriculture*. Oxford and IBH, New Delhi. pp. 91-112.

Partap U and Verma L R. 1994. Pollination of radish by *Apis cerana*. *Journal of Apicultural Research* 33: 237-241.

Partap U. 1997. *Bee flora of the Hindu Kush- Himalayas: Inventory and management*. ICIMOD, Kathmandu, Nepal.

Pascarella J B. 2008. *The Bees of Florida*. Department of Biology, Georgia Southern University.

Patiny S, Michez D and Danforth B N. 2007. Phylogenetic relationships and host-plant evolution within the basal clade of Halictidae (Hymenoptera, Apoidea). *Cladistics* 23: 17-21.

Paydas S, Eti S, Kaftanoglu O, Yasa E and Derin K. 2000. Effect of pollination of strawberries grown in plastic greenhouse by bumblebees on the yield and quality of the fruits. *Acta Horticulturae* 513: 443-451.

Peckolt T. 1893. Ueber brasilianische Bienen (as Melipona tataira and Trigona taga fogo [sic], folk names: "tataira", "caga-fogo"). *Nature* 42: 579-581.

Pedro S R M. 2014. The stingless bee fauna in Brazil (Hymenoptera: Apidae). *Sociobiology* 61(4): 348-354.

Pellmyr O. 2003.Yuccas, yucca moths, and coevolution: a review. *Annals Missouri Botany Garden* 90: 35–55.

Percy A P. 1989. Division of *Trigona iridipennis* Smith colony. *Indian Bee Journal* 51: 149.

Perez E, Rodríguez-Malaver A J and Vit P. 2006. Antioxidant capacity of Venezuelan honey in wistar rat homogenates. *Journal of Medical Food* 9(4): 510-516.

Phillips J P and Hilliker A J. 1990. Genetic analysis of oxygen defense mechanisms in *Drosophila melanogaster*. *Advances in Genetics* 28: 43–71.

Pick R A and Blochtein B. 2002. Atividade de coleta e origem floral do pólen armazenado em colônias de Plebeia saiqui (Holmberg) (Hymenoptera, Apidae, Meliponinae) no sul do Brasil. *Revista Brasileira De Zoologia* 19(1): 10.1590/S0101-81752002000100025.

Pinzauti M. 1993. Some observations on the bio-ethology, flight and foraging activity of *Osmia cornuta* Ltr, (Hymenoptera: Megachilidae) during strawberry pollination. *Apicultura* 8: 7-15.

Plowright R C and Laverty T M. 1984. Bumble bee and crop pollination in Ontario. *Proceeding of Entomological Society Ontario* 118: 155-160.

Pomeroy N and Fisher R M. 2002. Pollination of kiwi fruit (*Actinidia deliciosa*) by bumble bees (*Bombus terrestris*): effect of bee density and patterns of flower visitation. *New Zealand Journal of Entomology* 25: 41-49.

Ponchau O, Iserbyt S, Verhaeghe J C and Rasmont P. 2006. Is the caste ratio of the oligolectic bumblebee *Bombus gerstaeckeri* Morawitz (Hymenoptera: Apidae) biased to queens? *Annales de la Societe Entomologique de France* 42: 207–214.

Porporato M, Pinna M, Manino A and Marletto F. 1995. *Bombus terrestris* (L.) e *Apis mellifera* L. nell impollinazione del peperone in coltura protetta. *Modern Apiculture* 86: 99-112.

Portugal-Araújo V. 1955. *Colméias para abelhas sem ferrão*. Angola: Boletim do

Instituto de Angola. 31 p

Poulsen M H. 1973. The frequency and foraging behaviour of honeybees and bumble bees on field beans in Denmark. *Journal of Apicultural Research* 12(2): 75-80.

Prakash K B. 2002. *Pollination potentiality of Indian honeybee viz., Apis cerana on the production of cucumber (Cucumis sativus* (Linn.) S. W.: Cucurbitaceae). M Sc (Agri.) thesis, University of Agricultural Sciences, Bengaluru.

Premila K S, Devanesan S and Shailaja K K. 2013. Reduviid bug, *Acanthaspis siva* - A predator of honeybee and stingless bee in Kerala. *Insect Environment* 19: 3 p.

Punith H V. 2014. *Designing of suitable hives for rearing and multiplication of stingless bee, Tetragonula irridipennis (Smith)*. M Sc (Agri) thesis.

Quezada-Euan J J G and González-Acereto J. 1994. A preliminary study on the development of colonies of *Melipona beecheii* in traditional and rational hives. *Journal of Apicultural Research* 33: 167-170.

Quezada-Euan J J G and Gonzalez-Acereto J. 2002. Notes on the nest habits and host range of cleptobiotic *Lestrimelitta niitkib* (Ayala 1999) (Hymenoptera: Meliponini) from the Yucatan Peninsula, Mexico. *Acta Zoologica Mexicana* 86: 245–249.

Quezada-Euan J, May-Itza W and Gonzalez-Acereto J. 2001. Meliponiculture in Mexico: problems and perspective for development. *Bee World* 82(4):160–167.

Raakhee M and Devanesan S. 2000. Studies on behaviour of stingless bee, *Trigona iridipennis* Smith (Apidae: Meliponinae). *Indian Bee Journal* 62(4): 59-62.

Rahman A, Das P K, Rajkumari P, Saikia J and Sharmah D. 2015. Stingless Bees (Hymenoptera: Apidae: Meliponini): diversity and distribution in India. 4(1): 77-81.

Ramalho M, Giannini T C, Malagodibraga K S and Imperatriz-Fonseca V L. 1994. Pollen harvest by stingless bee foragers (Hymenoptera, Apidae, Meliponinae). *Grana* 33: 239–244.

Ramalho M, Imperatriz-Fonseca V L, Kleinert Giovannini A and Cortopassi-Laurino M. 1985. Exploitation of floral resources by *Plebeia remota* (Holmberg) (Apidae, Meliponinae). *Apidologie* 16: 307-330.

Ramalho M, Kleinert-Giovanni A and Imperatriz-Fonseca V L. 1991. *Ecologia alimentar de abelhas sociais*. In: J. R. Parra & A. R. Panizzi (org.). Ecologia nutricional de insetos e suas implicações no manejo de pragas. Ed. Manole & CNPq, São Paulo & Brasília.

Ramalho M, Kleinert-Giovannini A and Imperatriz-Fonseca V L. 1989. Utilization of floral resources by species of *Melipona* (Apidae, Meliponinae): floral preferences. *Apidologie* 20: 185–195.

Ramalho M, Kleinert-Giovannini A and Imperatriz-Fonseca V L. 1990. Important bee plants for stingless bees (Melipona and Trigonini) and Africanized honeybees (*Apis mellifera*) in neotropical habitats: a review. *Apidologie* 21: 469-488.

Ramalho M. 1990. Food harvesting by the stingless bee *Scaptotrigona* (Apidae, Meliponinae). *Journal of Apicultural Research* 29: 61-67.

Ramanujam C G K, Fatima K and Kalpana T P. 1993. Nectar and pollen sources for dammer bee (*Trigona iridipennis* Smith) in Hyderabad. *Indian Bee Journal*

55(1/2): 25-28.

Ramanujan C G K and Kalpana T P. 1992. *Tamarindus indica* L. an important forage plant for *Apis florea* F. in South Central India. *Apidology* 23: 403–413.

Ramirez-A E and Martínez H E. 2007. Melitopalynological characterization of *Scaptotrigona mexicana* Guérin (Apidae: Meliponini) and *Apis mellifera* L. (Apidae: Apini) honey Samples in Northern Puebla State, Mexico. *Journal of the Kansas Entomological Society* 80(4): 377-391.

Rangel J, Griffin S R and Seeley T D. 2010. Nest-site defense by competing honeybee swarms during house-hunting. *Ethology* 116: 608–618.

Rasmussen C. 2008. Catalog of the Indo-Malayan/Australasian stingless bees (Hymenoptera: Apidae: Meliponini). *Zootaxa* 1935: 1-80.

Rasmussen C. 2013. Stingless bee (Hymenoptera: Apidae: Meliponini) of the Indian subcontinent: Diversity, taxonomy and current status of knowledge. *Zootaxa* 3647(3): 401–428.

Rasmussen C. 2004. A stingless bee nesting with a paper wasp (Hymenoptera: Apidae, Vespidae). *Journal of the Kansas Entomological Society* 77: 593-601.

Ravestijin van W and Kraemer C. 1991. Hommel doet hetgoed op mini watermelon. *Groeten en Fruit* 46: 18-19.

Ravestijin van W and Sande J. 1991. Use of bumble bees for pollination of glasshouse tomatoes. *Acta Horticulturae* 288: 204-209.

Rayment T. 1935. *A cluster of bees*. Endeavour Press.

Reyes-Gonzalez A, Camou-Guerrero A, Reyes-Salas O, Arturo Argueta and Alejandro C. 2014. Diversity, local knowledge and use of stingless bees (Apidae: Meliponini) in the municipality of Nocupetaro, Michoacan, Mexico. *Journal of Ethnobiology and Ethnomedicine* 10(1): 47. DoI: 10.1186/1746-4269-10-4

Ribbands C R. 1949. The foraging method of individual honey-bees. *Journal of Animal Ecology* 18: 47–66.

Richard M F and Pomeroy N. 1989. Pollination of greenhouse muskmelon by bumble bees (Hymenoptera: Apidae). *Journal of Economic Entomology* 82: 1063-1066.

Richards K W and Myers T W. 1997. Commercially managed colonies of bumble bees for pollination of Cicer milkvetch. *Acta Horticulturae* 437: 293-296.

Robertson C. 1925. Heterotropic bees. *Ecology* 6: 412–436.

Rodriguez C D, Ramon-Sierra J, LopeAyora J, Sauri-Duch E, Cuevas-Glory L and Ortiz-Vazquez L. 2012. Antibacterial properties of honey produced by *Melipona beecheii* and *Apis mellifera* against food borne microorganisms. *Food Science Biotechnology* 21(3): 905-909.

Rodrigues C G, Krüger A P, Barbosa W F and Guedes R N C. 2016. Leaf fertilizers affect survival and behavior of the neotropical stingless bee *Friesella schrottkyi*. *Journal of Economic Entomology* 109(3): 1001-1008.

Ronsted N, Weiblen G D, Cook J M, Salamin N, Machado C A and Savolainen V. 2005. 60 million years of codivergence in the fig-wasp symbiosis. *Proceedings of the Royal Society of London Series B* 272: 2593–2599.

Roopa A N, Eswarappa G, Sanganna M Sajjanar and Gavi Gowda. 2015. Study on nesting characteristics and biology of stingless bee (*Trigona iridipennis* smith). *Journal of Agriculture and Veterinary Science* 8(10): 34-36.

Roowi S, Muhamad S A, Sipon H, Jaafar M F, Daud M N H and Othman R.

2012. Asid fenolik bebas dalam madu Kelulut. *Buletin Teknologi Mardi* 2: 145–147.

Roselino A C, Santos S B, Hrncir M and Bego L R. 2009. Differences between the quality of strawberries (Fragaria × ananassa) pollinated by the stingless bees *Scaptotrigona aff. depilis* and *Nannotrigona testaceicornis*. *Genetics and Molecular Research* 8(2): 539-545.

Roubik D and Moreno J E. 2009. *Trigona corvina*: An ecological study based on unusual nest structure and pollen analysis. *Psyche- A Journal of Entomology* 10: 295-314. DoI:10.1155/2009/268756.

Roubik D W and Moreno J E. 1990. *Social bees and palm trees: what do pollen diets tell us*? In: Social Insects and the Environment: Proceedings 11th International Congress. IUSSI, Bengaluru India. (Eds) GK Veeresh, B Mallik, C A Viraktamath, pp 427–28. EJ Brill Leiden.

Roubik D W, Moreno J E, Vergara C and Whittman D. 1986. Sporadic food competition with the African honeybee: projected impact on neoeotropical social bees. *Journal of Tropical Ecology* 2: 97-111.

Roubik D W, Weigt L A and Bonilla M A.1996. Population genetics, diploid males, and limits to social evolution of euglossine bees. *Evolution* 50: 931–935.

Roubik D W. 1982a. Obligate necrophagy in a social bee. *Science* 217: 1059–1060.

Roubik D W. 1983. Nest and colony characteristics of stingless bees from Panama (Hymenoptera: Apidae). *Journal of Kansas Entomological Society* 56: 327–355.

Roubik D W. 1989. Ecology and natural history of tropical bees, Cambridge Univ. Press, Cambridge, UK. *Journal of Tropical Ecology* 9: 248-248.

Roubik D W. 1995. *Pollination of cultivated plants in the tropics*. FAO Agriculture Service Bullettin No. 118. Food Agricultural Organization, Rome, Italy.

Roubik D W. 1995. *Stingless bee colonies for pollination*. (*In*) Roubik D W (Ed). Agriculture service bulletin, FAO (UN) No. 118, Rome, 324 p.

Roubik D W. 2006. Stingless bee nesting biology. *Apidologie* 37: 124–143.

Roubik D W. 2018. *The pollination of cultivated crops- a compendium for practitioners*, 2nd edition. FAO, Rome.

Roulston T H and Julier H E. 2009. Wild bee abumdance and pollination service in cultivated pumpkins: farm management, nesting behaviour and landscape effects. *Journal of Economic Entomology* 102(2): 563-573.

Sadeh A, Shmida A and Keasar T. 2007. The carpenter bee *Xylocopa pubescens* as an agricultural pollinator in greenhouses. *Apidologie* 38(6): 508–517.

Saini M S and Ghattor H S. 2007. Taxonomy and food plants of some bumble bee species of Lahaul and Spiti valley of Himachal Pradesh. *Zoo's Printing Journal* 22(4): 2648-2657.

Saini M S, Raina R H and Khan Z H. A check list of bumble bees (Hymenoptera: Apidae) from Indian Himalaya. *Journal of Insect Science* 24: 326-352.

Saint- Hilare A de. 1830. *Voyages dans les provinces de Rio De Jenerio et de Minas Geaeas*. 2: p.371.; 1848, Voyage aux sources de Rio de S. Francisco et dans la province de Goyaz. 2: 164.

Sakagami S F and Laroca S. 1963. Additional observations on the habits of the cleptobiotic stingless bees, the genus *Lestrimelitta friese* (Hymenoptera, Apoidea). *Journal of Factual Science Hokkaido University* 15: 319–339.

Sakagami S F, Inoue T, Yamane S and Salmah S. 1983. Nest architecture and colony composition of the Sumatran stingless bee *Trigona* (*Tetragona*) *laeviceps*. *Kontyu* 51(1): 100-111.

Sakagami S F, Inoue T and Salmah S. 1990. Stingless bees of central Sumatra. **In**: Sakagami S F, Ohgushi R and Roubik D W (Eds). Natural History of social wasps and bees in equatorial Sumatra, Hokkaido University Press, Sapporo pp 13-125.

Sakagami S F. 1978. Stingless bees (excl. *Tetragonula*) from the continental Asia and Sri Lanka (Hymenoptera: Apidae). *Journal of Science, Zoology Faculty* 21: 165-247.

Sakagami S F. 1982. Stingless bees. (*In*) Social Insects. H R Hermann (Ed). 3: 361–423.

Samoskorn J, Posri N and Pantong P. 2010. Efficacy of *Trigona pegdeni* for Pollination of F1 hybrid cucumber. *Plant Science* 13: 23-26.

Sanchez D, Nieh J C and Vandame R. 2011. Visual and chemical cues provide redundant information in the multimodal recruitment system of the stingless bee *Scaptotrigona mexicana* (Apidae, Meliponini). *Insectes Sociaux* 58: 575-579.

Sanchez L A, Picado A, Sommeijer M J and Slaa E J. 2002. Floral biology, pollination ecology and seed production of the ornamental plant Salvia splendens Sello. *Journal of Horticultural Science and Biotechnology* 77: 498–501.

Santos C F, Acosta A L, Nunes-Silva P, Saraiva A M and Blochtein B. 2015. Climate warming may threaten reproductive diapause of a highly eusocial bee. *Environmental Entomology* 44: 1172–1181.

Santos C F, Santos P D and Blochtein B. 2016. *In vitro* rearing of stingless bee queens and their acceptance rate into colonies. *Apidologie* 47: 539–547.

Santos dos C G, Megiolaro F L, Serrao J E and Blochtein B. 2009. Morphology of the head salivary and intramandibular glands of the stingless bee *Plebeia emerina* (Hymenoptera: Meliponini) workers associated with propolis. *Annals of Entomological Society of America* 102:137-143.

Santos Junior M C and Santos F A R. 2003. Espectro Polinico de amostras de meis coletdas na microreggiao do papagussau Bahia. *Magistra* 24:158-171.

Santos S A B dos, Bego L R and Roselino A C. 2004. *Pollination in tomatoes, Lycopersicon esculentum, by Melipona quadrifasciata and Apis mellifera (Hymenoptera, Apinae)*. (*In*) Proc. 8th IBRA Int. Conf. Trop. Bees and VI Encontrosobre Abelhas, p 688.

Santos S A B, Roselino A C, Hrncir M and Bego L R. 2009. Pollination of tomatoes by the stingless bee *Melipona quadrifasciata* and the honeybee *Apis mellifera* (Hymenoptera, Apidae). *Genetic Molecular Research* 8: 751–757.

Sanz J M G and Serrano A R. 2006. Quality fruit improvement in sweet pepper culture by bumble bee pollination. *Scientia Horticulturae* 110(2): 160-166.

Sarto M C L del, Peruquetti R C and Campos L A O. 2005. Evaluation of the Neotropical stingless bee *Melipona quadrifasciata* (Hymenoptera: Apidae) as pollinator of greenhouse tomatoes. *Journal of Economic Entomology* 98: 260-266.

Sawatthum A. 2017. Efficacy of stingless bee *Lepidotrigona terminata* as insect pollinator of F_1 hybrid cucumber. *International Journal of Geomate*. DoI -

10.21660/2017.37.2533

Schmidt V M, Zucchi R and Barth F G. 2003. A stingless bee marks the feeding site in addition to the scent path (*Scaptotrigona* aff. *deplis*). *Apidologie* 34: 237–248.

Schultz W A. 1907. Zeistchr Wiss Insekten. *Biology* 3: 67.

Schwarz H F. 1948. Stingless bees (Meliponidae) of the western hemisphere. *Lestrimellita* and the following subgenera of *Trigona: Trigona, Paratrigona, Schwarziana, Parapartamona, Cephalotrigona, Oxitrigona, Scaura* and *Mourella*. *Bullettin of American Museum of Natural History* 90: 1-546.

Schwarz H F. 1939. The Indo-Malayan species of *Trigona*. *Bulletin of the American Museum of Natural History* 76: 83-141.

Seeley T D. 1983. Division of labor between scouts and recruits in honeybee foraging. *Behavioural Ecology and Sociobiology* 12: 253–259.

Seeley T D. 1985. *Honeybee ecology.* Princeton University Press, Princeton, New Jersey.

Seeley T D. 1985. *The information-center strategy of honeybee foraging*. (*In*) Hölldobler B, Lindauer M (Eds). *Experimental Behavioral Ecology.* G. Fischer Verlag, New York, pp 75–90.

Seeley T D. 1989a. The honeybee colony as a superorganism. *American Science* 77: 546–553.

Seeley T D. 1989b. Social foraging in honeybees: how nectar foragers assess their colony's nutritional status. *Behavioural Ecology and Sociobiology* 24: 181–199.

Shanks J L, Haigh A M, Riegler M and Hart Robert N S. 2017. First confirmed report of a bacterial brood disease in stingless bees. *Journal of Invertebrate Pathology* 144: 7-10.

Shanks J L. 2015. *Tetragonula carbonaria and disease: behavioural and antimicrobial defences used by colonies to limit brood pathogens*. Ph D thesis, School of Science and Health, Western Sydney University.

Sharma D, Katna S, Sharma R, Rana B S, Sharma H K, Bhardwaj V and Chauhan A. 2019. First detection of *Nosema Ceranae* Infecting *Apis Mellifera* in India. *Journal of Apicultural Science* 63. 10.2478/jas-2019-0002.

Sharma H K and Gupta J K. 1997. Pollen analysis of nectar and honey sac contents of *Apis cerana* and *Apis mellifera* at Katrain during different seasons. Paper presented at the National Conference on Apiculture, Bengaluru.

Sheffield C S, Hebert P D N, Kevan P G and Packer L. 2009. DNA barcoding a regional bee (Hymenoptera: Apoidea) fauna and its potential for ecological studies. *Molecular Ecology Resources* 9: 196-207.

Shipp J L, Whitefield G H and Papadapolous A P. 1994. Effectiveness of bumble bee, *Bombus impatiens* Cr. (Hymenoptera: Apidae), as a pollinator of greenhouse sweet pepper. *Scientia Horticulturae* 57: 29-39.

Sihag R C and Rathi A. 1994. Diversity, abundance, foraging behavior and pollinating efficiency of different bees visiting pegion pea (*Cajanus cajan*) blossoms. *Indian Bee Journal* 56(3-4): 187-201.

Silva A C and Gl-Santana H R. 2004. Predation of *Apiomerus pillipes* (Fab.) (Hemiptera, Reduviidae, Harpactorinae, Apiomerini) over meliponinae bees (Hymenoptera, Apidae) in the state of Amazonas, Onas, Brazil. *Revista Brasileria de Zoologia* 21: 769- 774.

Silva C I, Maia-Silva C, Ribeiro F A and Bauermann S G. 2012. *O uso da palinologia como ferramenta em estudos sobre ecologia e conservação de polinizadores no Brasil*. pp. 369–383. In Imperatriz-Fonseca V, Canhos D, Alves D, Saraiva A (Eds). Polinizadores no Brasil - Contribuição e Perspectivas para a Biodiversidade, Uso Sustentável, Conservação e Serviços Ambientais. EDUSP; São Paulo, Brasil 488 pp.

Silveira F A, Melo G A R and Almeida E A B. 2002. *Abelhas Brasileiras: Sistemática e Identificação*. Fernando A. Silveira p 253.

Simoes D, Bego L R, Zucchi R, and Sakagami S F. 1980. *Melalonctia sinistra* an endoparasite phorid fly attacking *Nannotrigona (Scaptotrigona) postica* (Hymenoptera: Meliponinae). *Revista Brazilia de Entomologica* 24: 137-142.

Simone F M and Spivak M. 2010. Propolis and bee health: the natural history and significance of resin use by honeybees. *Apidologie* 41: 295-311.

Singh A K. 2016. Traditional meliponiculture by Naga Tribe in India. *Indian Journal of Traditional Knowledge* 15(4): 693-699.

Singh A K, Jaiswal D K, Singh H K and Thakur H K. 2016. Diversity of bees, flora and floral calendar of native honeybees in Nagaland, India. *Advances in Life Sciences* 5(6): 2285-2292.

Singh M P. 1999. Melissopalypalynology and survey of the bee plants in northeastern hill region. *Indian Journal of Hill Farming* 12(2): 79-83.

Siqueira K M M de, Kill L H P, Martins C F, Lemos I B, Monteiro S P and Feitoza E A. 2009. Ecology of pollination of the yellow passion fruit (*Passiflora edulis* sims f. flavicarpa deg.), in the region of Sao Francisco Valley. *Revista Brasileira de Fruticultura* 31(1): 1–12.

Slaa E J, Sanchez L A C, Malagodi-Braga K S and Hofstede F E. 2006. Stingless bees in applied pollination: practice and perspectives. *Apidologie* 37: 293-315.

Slaa E J, Wassenberg J and Biesmeijer J C. 2003. The use of field-based social information in eusocial foragers: local enhancement among nestmates and heterospecifics in stingless bees. *Ecological Entomology* 28: 369–379.

Smiley R L, Baker E W and Delfinado M. 1996. New species of *Hypoaspis* (Acari: Mesostigmata: Laelapidae) from the nest of stingless bee in Malaysia (Hymenoptera: Meliponinae, Apidae). *Serial Zoology* 67(2): 197-203.

Smith D D. 1998. Iowa prairie: original extent and loss, preservation and recovery attempts. *Journal of Iowa Academy of Science* 105: 94-108.

Smith F G. 1960. *Beekeeping in the tropics*. Longman & Green & Co Ltd. No.-8 3/4 × 5 1/2, pp. + 265 p.

Soares H M, Jacob C R O, Carvalho S M, Nocelli R C F and Malaspina O. 2015. Toxicity of imidacloprid to the stingless bee *Scaptotrigona postica* Latreille, 1807 (Hymenoptera: Apidae). *Bullettin of Environment Contamination and Toxicology* 1807: 675-680. DoI: 10.1007/ s00128-015-1488-6

Sommeijer M J, de Rooy G A, Punt W and de Bruijn L L M. 1983. A comparative study of foraging behavior and pollen resources of various stingless bees (Hym., Meliponinae) and honeybees (Hym., Apinae) in Trinidad, West-Indies. *Apidologie* 14: 205–224.

Sommeijir M J. 1999. Beekeeping with stingless bees: a new type of hive. *Bee World* 80(2): 70-79.

Sommerlandt F M J, Huber W and Spaethe J. 2014. Social Information in the stingless bee, *Trigona corvina* Cockerell (Hymenoptera: Apidae): the use of visual and olfactory cues at the food site. *Sociobiology* 61(4): 401-406.

Sousa L A, Pereira T O, Prezoto F and Faria M G M. 2002. Nest foundation and diversity of Meliponini (Hymenoptera, Apidae) in an urban area of the municipality of Juiz de Fora, MG, Brasil. *BioScience Journal* 18: 59-65.

Souza B, Roubik D, Barth O, Heard T, Enriquez E, Carvalho C, Villas-Booas J, Almeida-Muraidan L, Bogdanov S and Vit P. 2006. Composition of stingless bee honey: setting quality standards. *Interciencia* 31: 867-875.

Spooner-Hart R, Halcroft M and Shanks J. 2018. *Presentation on stingless bee pests and diseases*. Western Sydney University, Australia.

Stanghellini M S, Ambrose J T and Schultheis J R. 1997. The effects of honeybee and bumble bee pollination on fruit set and abortion of cucumber and watermelon. *American Bee Journal* 137: 386-391

Stanghellini M S, Ambrose J T and Schulthies J R. 1998a. Seed production in watermelon: a comparison between two commercially available pollinators. *Horticultural Science* 33: 28-30.

Stanghellini M S, Ambrose J T and Schulthies J R. 1998b. Using commercial bumble bee colonies as backup pollinators for honeybees to produce cucumbers and watermelon. *Horticultural Technology* 8: 590-594.

Stanghellini M S, Ambrose J T and Schulthies J R. 2002. Diurnal activity, floral visitation and pollen deposition by honeybees and bumble bees on filed grown cucumber and watermelon. *Journal of Apicultural Research* 40(1/2): 27-34.

Stubbs C S and Drummond F A. 2001. *Bombus impatiens* (Hymenoptera: Apidae): an alternative to *Apis mellifera* (Hymenoptera: Apidae) for lowbush blue berry pollination. *Journal of Economic Entomology* 94(3): 609-616.

Suryanarayana M C. 1966. Studies on flora of Coorg with emphasis on beekeeping forage plants, ecological and botanical features. *Indian Bee Journal* 28: 59-75.

Suwannapong G, Eiri D and Benbow M E. 2012. Honeybee communication and pollination, pp 1-24.

Syafrizal T D and Yusuf R. 2014. Biodiversity and habitat of trigona at secondary Tropical Rain Forest of Lempake Education Forest, Samarinda, Kalimantan Timur. *Journal Teknologi Pertanian* 9(1): 34-38.

Tej M K, Srinivasan M R, Rajashree V and Thakur R K. 2017. Stingless bee *Tetragonula iridipennis* Smith for pollination of greenhouse cucumber. *Journal of Entomology and Zoology Studies* 5(4): 1729-1733.

Temaru E, Shimura S, Amano K and Karasawa T. 2007. Antibacterial activity of honey from stingless honeybees (Hymenoptera; Apidae; Meliponinae). *Polish Journal of Microbiology* 56: 281-285.

Terada Y. 1972. *Enxameagem em Frieseomelitta varia Lepeletier (Hymenoptera: Apidae)*. (*In*) Homenagem à W. E. Kerr. Rio Claro, SP Brasil, pp 293–299.

Thakur R K and Kashyap L. 2008. *Record of rearing bumble bee, Bombus haemorroidalis Smith in captivity and their utilization in polyhouse crop pollination*. 2nd International Beekeeping Congress, August 19-21. Thimpu, Bhutan. pp. 31-33.

Thakur R K, Chauhan A, Sharma M and Gupta R K. 2010. Impact of bumble bee pollination on cucumber productivity under protected conditions. *Pest Management and Economic Zoology* 18(1/2): 19-25.

Thakur R K, Gupta J K and Gupta P R. 2005. *Investigation on rearing of bumble bees (Bombus spp.) in captivity*. (*In*) *International Beekeeping Congress*, November

13th to 18th, Century foundation in association with Mountain Research Development Associates (MRDA), FAO, Bangalore, India, 18 p.

Thakur R K. 2006. *Bumble bee flora and domestication of bumble bee (Bombus sp.) under midhill conditions of Himachal Pradesh*. (*In*) International Workshop on integrated beekeeping development in South Asian countries, Nov-13-14, 2006, New Delhi, p 30.

Thakur R K. 2011. *Bumble bee rearing technology and its utilization in pollinating crops*. Dr YSP University of Horticulture and Forestry, Nauni, Solan, Himachal Pradesh.

Thakur S. 2018. *Studies on bumble bee (Bombus haemorrhoidalis Smith) pollination in Capsicum annuum L. under protected cultivation*. M Sc thesis, Dr Y S Parmar University of Horticulture and Forestry, Nauni, Solan.

Thies S A. 1953. Agents concerned with natural crossing of cotton in Oklahoma. *Agronomy Journal* 45: 481-484.

Thompson J D and Goodell K. 2001. Pollen removal and deposition by honeybees and bumble bees visitors to apple and almond flowers. *Journal of Applied Ecology*. 38: 1032-1044.

Tome H V V, Martins G F, Lima M A P, Campos L A O and Guedes R N C. 2012. Imidacloprid-induced impairment of mushroom bodies and behavior of the native stingless bee *Melipona quadrifasciata anthidioides* 7(6): 38406. https://DoI.org/10.1371/journal.pone.

Tome H V V, Barbosa W F, Correa A S, Gontijo L M, Martins G F and Guedes R N C. 2015. Reduced-risk insecticides in Neotropical stingless bee species: Impact on survival and activity. *Annals of Applied Biology* 167(2): 186-196.

Trelease W M. 1879. *Nectar- what it is, and some of its uses*. USDA, pp 319-343.

Tung C C and Kong H F. 1997. Evaluation of the pollination efficiency of bumble bee (*Bombus terrestris*) on green house tomatoes. *Chinese Journal of Entomology* 16(3): 167-175.

Valdovinos-Nunez G R, Quezada-Euan J J G, Ancona-Xiu P, Moo-Valle H, Carmona A and Ruiz Sanchez E. 2009. Comparative toxicity of pesticides to stingless bees (Hymenoptera: Apidae: Meliponini). *Journal of Economic Entomology* 102(5): 1737-1742. DoI: 10.1603/029.102.0502

Valle A A, Aramayo E, Gil M and Lamberto S. 2007. A melissopalynological map of the south and southwest of the Buenos Aires Province, Argentina A. *Spanish Journal of Agricultural Research* 5(2): 172-180.

Veen J W van, Sommeijer M J and Meeuwsen F. 1997. Behaviour of drones in *Melipona* (Apidae, Meliponinae). *Insectes Sociaux* 44: 435–447.

Veen J W van and Sommeijer M J. 2000. Colony reproduction in *Tetragonisca angustula* (Apidae, Meliponini). *Insectes Sociaux* 47: 70–75.

Velthius H W and Laura C. 1991. Pollination of primula in a greenhouse using bumble bees. *Acta Horticulturae* 288: 199-203.

Velthuis H H W and Doorn A-Von. 2006. A century of advances in bumble be keeping and the economic and environmental aspects of its commercialization for pollination. *Apidologie* 37(4): 421-451.

Velthuis H H W and Sommeijer M J. 1991. *Roles of morphogenetic hormones in caste polymorphism in stingless bees*. (*In*) Gupta AP (ed). *Morphogenetic hormones of arthropods.* Rutgers University Press, Rutgers, NJ, pp 346–383.

Venturieri G C, Alves D A, Villas-Boas J K, Carvalho C A L, Menezes C and Vollet-

Neto A. 2012. *Meliponicultura no Brasil: situação atual e perspectivas futuras.* (*In*) Imperatriz-Fonseca V L, Canhos D A L, Alves D A, Saraiva A M (Eds). Polinizadores no Brasil contribuição e perspectivas para a biodiversidade, uso sustentável, conservação e serviços ambientais. São Paulo: EDUSP.

Venturieri G C, Raiol V F O and Pereira C A B. 2003. Avaliação da introdução da criação racional de Melipona fasciculata (Apidae: Meliponina), entre os agricultores familiares de Bragança - PA, Brasil. *Biota Neotropica* 3: 1-7.

Venturieri G C. 2008a. Caixa para a criação de UruçuAmarela *Melipona flavolineata* Friese, 1900. *Comunicado Técnico Embrapa Amazônia Oriental.* 212: 1-8.

Viana B F and Kleinert A M P. 2005. A community of flower-visiting bees (Hymenoptera: Apoidea) in the coastal sand dunes of north-eastern Brazil. *Biota Neotropica* 5(2): 1-13.

Vijayakumar K, Muthuraman M and Jayaraj R. 2013. Infestation of *Carpoglyphus lacti* (Linnaeus) (Acari: Carpoglyphidae) on *Trigona iridipennis* (Apidae: Meliponinae) from India. *Journal of Agricultural Sciences* 3(1): 25-28.

Vijayan M, Saravanan P A and Srinivasan M R. 2018. Effect of season and timings on the foraging behaviour of stingless bee, *Tetragonula iridipennis* Smith (Hymenoptera: Apidae). *Madras Agriculture Journal* 105(7-9): 286-290.

Villanueva-G R, Roubik D W and Colli-Ucán. 2005. Extinction of *Melipona beecheii* and traditional beekeeping in the Yucatán peninsula. *Bee World* 86: 35-41.

Viraktamath S, Lingappa S, Kallesha G R, Holi K S and Hunur S. 1999. Comparative studies on pollen sources to *Trigona* sp., *Apis mellifera* L. and *Apis cerana* F. in Dharwad, Karnataka, India. *Indian Bee Journal* 61: 71–73.

Visscher P K and Seeley T D. 1982. Foraging strategy of honeybee colonies in a temperate deciduous forest. *Ecology* 63: 1790-1801.

Vit P, Medina M and Enriquez E. 2004. Quality standards for medicinal uses of Meliponinae honey in Guatemala, Mexico and Venezuela. *Bee World* 85(1): 25-31.

Vit P, Pedro S R and Roubik D. 2013. *Pot-honey: a legacy of stingless bees.* Springer, New York, pp 219–227.

Vit P, Ricciardelli D and Albore G. 1994. Melissopalynology for stingless bees (Hymenoptera: Apidae: Meliponinae) in Venezuela. *Journal of Apicultural Research* 33: 145-154.

von Frisch K. 1967. *The dance language and orientation of bees,* 2nd printing. 1993. Belknap Press, Cambridge, Mass.

Vossler F G, Blettler D C, Fagúndez G A and Dalmazzo M. 2018. *Stingless bees as potential pollinators in agroecosystems in Argentina: Inferences from Pot-Pollen studies in natural environments.* (*In*) Vit P, Pedro S, Roubik D (eds). Pot-Pollen in stingless bee Melittology, Springer, Cham, New York.

Vossler F G, Tellería M C and Cunningham M. 2010. Floral resources foraged by *Geotrigona argentina* (Apidae, Meliponini) in the Argentine Dry Chaco forest. *Grana* 49(2): 142-153. DoI: 10.1080/00173131003694274

Waller G D, Vissiere B E, Moffett J O and Martin J H. 1985. Comparison of carpenter bees (*Xylocopa varipuncta* Patton) (Hymenoptera: Anthophoridae) and honeybees (*Apis mellifera* L.) (Hymenoptera: Apidae) as pollinators of male-sterile cotton in cages. *Journal of Economic Entomology* 78: 558–561.

Wappler T, Labandeira C C, Engel M S, Zetter R and Grimsson F. 2015. Specialized and generalized pollen-collection strategies in an ancient bee lineage. *Current Biology* 25:1-7.

Waser N and Ollerton J. 2006. *Plant-pollinator interactions: from specialization to generalization*. Bibliovault OAI Repository, University of Chicago Press.

Waser N M, Chittka L, Price M V, Williams N M and Ollerton J. 1996. Generalization in pollination systems, and why it matters. *Ecology* 77(4): 1043-1060.

Watmouth R H. 1974. Biology and behavior of carpenter bees in southern Africa. *Journal of the Entomological Society of South Africa* 37: 261–281.

Wattena W and Jongjitvimol T. 2007. First record of the predator, *Pahabengkakia piliceps* Miller, 1941 (Reduviidae, Harpactorinae) in the Stingless Bee, *Trigona collina* Smith, 1857(Apidae, Meliponinae) in Thailand. *Natural History Journal of Chulalongkorn University* 7: 71- 74.

Wcislo W T and Cane J H. 1996. Floral resource utilization by solitary bees (Hymenoptera: Apoidea) and exploitation of their stored foods by natural enemies. *Annual Review of Entomology* 41: 257–286.

Weaver J B. 1978. *Observations on bee activity in several genotypes of cotton*. USA, pp 76-77.

Westrich P. 1989. *Die Wildbienen Baden- Württembergs*. Stuttgart, Ulmer.

Whidden T L. 1996. The fidelity of commercially reared colonies of *Bombus impatiens* Cresson (Hymenoptera: Apidae) to lowbush blueberry in Southern New Brunswick. *Canadian Entomologist* 128: 957-958.

Witter S and Tirelli F. 2014. *Polinizadores nativos presentes em Lavouras de Canola no Rio Grande do Sul*. In: abelhas na polinizacao da canola. IPUCRS, Porto Alegre, Brasil.

Whittington R and Winston M L. 2004. Comparison and examination of *Bombus accidentalis* and *Bombus impatiens* (Hymenoptera: Apidae) in tomato greenhouses. *Journal of Economic Entomology* 97(4): 1384-1389.

Wille A .1983. Biology of the stingless bees. *Annual Review of Entomology* 28: 41–64.

Wille A and Orozco E. 1975. Observations on the founding of a new colony by *Trigona cupira* (Hymenoptera: Apidae) in Costa Rica. *Revista Tropical Biologica* 22: 253–287.

Wille A. 1965. Las abejas atarr´a de la regi´on mesoamericana del g´enero y subg´enero Trigona (Apidae-Meliponini). *Revista de Tropical Biologica* 13: 271–291.

Wille A. 1979. Phylogeny and relationships among the genera and subgenera of the stingless bees (Meliponinae) of the world. *Review Tropical Biology* 27(2): 241-277.

Williams P H. 1998. An annotated checklist of bumble bees with an analysis of patterns of description (Hymenoptera: Apidae, Bombini). *Bulletin of the Natural History Museum, Entomology* 67: 79-152.

Willmer P G, Bataw A A M and Hughes J P. 1994. The superiority of bumble bees to honeybees as pollinators: insect visits to raspberry flowers. *Ecological Entomology* 19: 271-284.

Wilms J and Eltz T. 2008. Foraging scent marks of bumblebees: footprint cues rather than pheromone signals. *Naturwissenschaften* 95: 149-153.

Winston M L and Michener C D. 1977. Dual Origin of highly social behavior among bees. *Proceedings of Natural Academic Science Washington* 74: 1135-1137.

Winter K, Adams L, Thorp R W, Inouye D W, Day L, Ascher J and Buchmann S. 2006. *Importation of non-native bumble bees into North American potential consequences of using Bombus terrestris and other non-native bumble bees for greenhouse crop pollination in Canada, Mexico, and the United States*. North American Pollinator Protection Campaign, San Fransisco, CA, p 33.

Wolf S and Moritz R F A. 2008. Foraging distance in *Bombus terrestris* L. (Hymenoptera: Apidae). *Apidologie* 39: 419-427.

Woo- kunsuk, Kimjhiwan S A, Mappatoba S A, Woo K S and Kim J H. 1996. The foraging activity of stingless bees, *Trigona* sp. (Apidae: Meliponinae) in the greenhouse. *Korean Journal of Apiculture* 11(2): 82-88.

Wunderink M E and Sommeijer M J. 1996. The ultimate function of the positioning and ovipositioning process (POP) in stingless bees: an experimental study. *Proceedings of Society of Experimental and Applied Entomology* 7: 51-55.

Yacoob M, Rajab N F, Shahar and Sharif R. 2017. Stingless bee honey and its potential value- a review. *Food Research* 2(2): 124-133.

Yankit P, Rana K, Sharma H K, Thakur M and Thakur R K. 2018. Effect of bumble bee pollination on quality and yield of tomato (*Solanum lycopersicum* mill.) grown under protected conditions. *International Journal of Current Microbiology and Applied Sciences* 7(1): 257-263.

Yazan L S, Muhamad Zali M F S, Ali R M, Zainal N A, Esa N, Sapuan S and Syed Alwi S S. 2016. Chemopreventive properties and toxicity of kelulut honey in sprague dawley rats induced with azoxymethane. *BioMed Research International*. DoI: 10.1155/2016/4036926.

Yokoi T and Fujisaki K. 2011. To forage or not: Responses of bees to the presence of other bees on flowers. *Annual Entomological Society of America* 104: 353-357. DoI: 10.1603/an10053

Zainol M I, Yusoff K M and Mohd Yusof M Y. 2013. Antibacterial activity of selected Malaysian honey. *B M C Complementary and Alternative Medicine* 13: 129.

Zeil J and Wittmann D. 1993. Landmark orientation during the approach to the nest in the stingless bee *Trigona (Tetragonisca) angustula* (Apidae, Meliponinae). *Insectes Sociaux* 40: 381-389.

APPENDIX 1

Fauna Included in the Book

Table 1. List of animal fauna included in the book

Apis mellifera	Honeybees
Apis cerana	
Apis florea	
Apis dorsata	
Tetragonula iridipennis	Stingless bees
Tetragonula laeviceps	
Tetragonula carbonaria	
Tetragonula bengalensis	
Tetragonula gressitti	
Tetragonula praeterita	
Tetragonula ruficornis	
Tetragonisca angustula	
Trigonisca pipioli	
Trigona mellipes	
Trigona collina	
Trigona amalthea	
Trigona hockingsi	
Trigona clypearis	
Mellipona scutellaris	
Melipona quadrifasciata	
Melipona rufiventris	
Melipona compressipes manaosensis	
Melipona seminigra	
Melipona subnitida	
Meliponula ferruginea	
Dactyrulina schmidti	
Plebeina hildebrandti	
Meliponula lendliana	
Meliponula bocandei	
Lepidotrigona ventralis	
Lepidotrigona arcifera	
Lepidotrigona flavibasis	
Lophotrigona canifrons	
Melipona beechi	
Lisotrigona cacciae	

(*Continued*)

Table 1. (*Continued*)

Lisotrigona mohandasi	
Scaptotrigona polistycta	
Scaptotrigona depilis	
Austroplebeia australis	
Partamona bilineata	
Plebeia fulvopilosa	
Plebia schrottkyi	
Plebeia saiqui	
Nannotrigona perilampiodes	
Nannotrigona testaceicornis	
Geotrigona acapulconis	
Geotrigona argentina	
Hypotrigona gribodoi	
Lestrimelitta chamelensis	
Partamona helleri	
Frieseomelitta varia	
Trigona buyssoni	Stingless bees
Lestrimelitta limao	
Bombus haemorrhoidalis	Bumble bees
B. kereinsis	
B. himalayanus	
B. asiaticus	
B. personatus	
B. rufofaciatus	
B. trifaciatus	
B. tumicatus	
B. consobrinus	
B. gerstaeckeri	
B. pascorum	
B. terrestris	
B. pratorum	
B. hyponorum	
B. accidentalis	
B. pennysylavanicus	
B. assmensis	
Xylocopa fenestrata	Carpenter bees
X. leucothorax	
X. tenuiscapa	
X. varipuncta	
Halictus lucidipennis	Halictid bees
Sphecodes sp.	
Lasioglossum sp.	
Ceriana ornata australis	Syrphid fly
Dohrniphora trigonae	Phorid fly
Megaselia scalaris	
Plega hagenella	Mantis flies

(*Continued*)

Table 1. (*Continued*)

Bembix musca	Sand wasp
Syntretus trigonaphagus	Braconid wasp
Pyemotes tritici	Grain itch mite
Aethina tumida	Beetle
Hermetia illucens	Phorid flies
Mealonctia sinistia	
Megaselia scalaris	
Rhizoplatys mucronatus	Beetles
Carpophilus spp.	
Brachypeplus spp.	
Megachile disjuncta	Megachilids
Megachile lanatus	
Acanthaspis siva	Bugs
Sycanus sp.	
Apiomerus pillipes	
Pahabengkakia piliceps	
Neotydeolus therapeutikos	Mite
Meliponopus palpifer	
Neocypholaelaps phooni	
Eumellipites sp.	
Amblyseius sp.	
Carpoglyphus lactis	
Pyemotes tritici	
Thomisus projectus	Spiders
Crossopriza lyoni	
Heteropoda venatoria	
Hersilia savignyi	
Plexippus paykulli	
Menemerus bivittatus	
Neoscona nautica	
Thomisus lobosus	
Nephila clavipes	
Nephila maculate	
Solenopsis geminata	Ants
Oecophylla smaragdina	
Elminia longicauda teresita	Blue fly catcher bird
Paradoxurus hermaphrodites	Mammals
Melursus ursinus	
Mellivora capensis	
Pan troglodytes	
Eira barbara	
Tamandua tetradactyla	
Lysinibacillus sphaericus	Bacteria
Paecilomyces variotii	Fungi
Aspergillus niger	

■

Index